PENGUIN BOOKS

STARTUP

Jerry Kaplan hasn't learned his lesson. Not long after GO was sold to AT&T, he started a new company—ON-SALE—devoted to reinventing electronic shopping. He lives near San Francisco, but can be found on the World Wide Web at http://www.onsale.com.

STARTUP

A Silicon Valley Adventure

JERRY KAPLAN

PENGUIN BOOKS

For Lily Layne Kaplan
Born August 28, 1994

PENGUIN BOOKS
Published by the Penguin Group
Penguin Group (USA) Inc., 375 Hudson Street, New York, New York 10014, U.S.A.
Penguin Books Ltd, 80 Strand, London WC2R 0RL, England
Penguin Books Australia Ltd, 250 Camberwell Road, Camberwell, Victoria 3124, Australia
Penguin Books Canada Ltd, 10 Alcorn Avenue, Toronto, Ontario, Canada M4V 3B2
Penguin Books India (P) Ltd, 11 Community Centre, Panchsheel Park, New Delhi – 110 017, India
Penguin Books (N.Z.) Ltd, Cnr Rosedale and Airborne Roads, Albany, Auckland, New Zealand
Penguin Books (South Africa) (Pty) Ltd, 24 Sturdee Avenue,
Rosebank, Johannesburg 2196, South Africa

Penguin Books Ltd, Registered Offices: 80 Strand, London WC2R 0RL, England

First published in the United States of America by Houghton Mifflin Company 1994
Reprinted by arrangement with Houghton Mifflin Company
Published in Penguin Books 1996

9   10

THE LIBRARY OF CONGRESS HAS CATALOGUED THE HARDCOVER AS FOLLOWS:
Kaplan, Jerry.
Startup: a Silicon Valley adventure/Jerry Kaplan.
p.   cm.
Includes index.
ISBN 0-395-71133-9 (hc.)
ISBN 0 14 02.5731 4 (pbk.)
1. GO Corporation—History.   2. Computer industry—California—San Mateo
County.   3. Entrepreneurship—United States—Case studies.   4. Pen-based
computers—History.   5. Kaplan, Jerry (Samuel Jerrold Kaplan).   I. Title.
HD9696.C64G615   1995
338.7'6100416—dc20   94–45110

Printed in the United States of America
Set in Granjon
Designed by Melodie Wertelet

CONTENTS

Prologue 1

1. The Idea 4
2. The Deal 21
3. The Company 37
4. The Financing 59
5. The Customer 82
6. The Proposal 97
7. The Partner 117
8. The Announcement 146
9. The War 170
10. The Spinout 196
11. The Switch 210
12. The Bubble 235
13. The Reversal 250
14. The Showdown 268

Epilogue 291

Author's Note 301
Chronology 303
Appendix 305
Glossary 307
Index 313

Prologue

GOING, GOING, GONE. The auction was over. The last of the obsolete personal computers, engineers' cubicles, and other debris of a corporate shipwreck was finally liquidated, sold piecemeal to a crowd of hopeful entrepreneurs looking for a bargain to help float their new ventures. A few curious bottom fishers hovered around the stacked remains of electronic pens, flat-panel displays, and plastic cases, picking over the artifacts of the dead company's product: a portable computer operated by a pen instead of a keyboard.

To those of us who had pinned our hopes on this novel concept, the auction seemed vaguely sacrilegious, like watching treasure hunters dredge up human remains in their search for valuables. But it was clear to me, as the person who had launched the enterprise in the first place, that our passions and ideas had simply outlived their host, only to take root elsewhere in Silicon Valley. GO Corporation, and its offspring, EO, never quite found its market, but the concept of a pen computer remains as seductive as ever.

Still, I had to accept that impossible, final truth: GO was gone. Six years, hundreds of jobs, $75 million — all gone. If statistics were all that mattered, the story would end here. But behind the numbers lies a portrait of life at the edge of the corporate universe, where the intrepid and the imprudent play a perpetual high-stakes game of creation. The goal is to establish new companies, magical engines of prosperity that spawn products, jobs, and wealth. The price of admission is a radical idea, one powerful enough to motivate people, attract investment, and focus society's energy on improving the way people work and play. But there is also a darker side to the story, a cautionary tale about what can happen to a young company when its timing is wrong, its technology too speculative, and its market not yet ready.

▣

As the winning bidders arranged to pick up their goods, I realized that the origin of GO could be traced back well before its founding in 1987, to a day in early 1979 when I first learned the truth about scientific progress from my Ph.D. dissertation advisor at the University of Pennsylvania.

A shy Indian man with a shiny, balding head and an occasional stutter, Dr. Joshi was widely known for his brilliant work in artificial intelligence. Our weekly meetings to help me find a thesis topic were more like therapy sessions than academic discussions. Most of the time he would sit silently behind his desk, watching me wrestle with some difficult question at the blackboard. When I was particularly down, he would offer a cryptic bit of encouragement: "You're not wrong, you know."

I had spent the past several months puzzling obsessively over an obscure problem in computational linguistics. One day, I explained to Dr. Joshi that I had searched the entire library for a clue to the solution, but without success.

"Perhaps you should try a different approach, Jerry."

"Like what?"

He pointed to the clock on his wall. It was round with no numerals, only single tick marks for the hours. "What time is it?"

"Four-thirty." I thought he was pointing out that our hour was up. Instead, he walked over and rotated the clock a quarter turn to the right.

"Now what time is it?" In its new position, the clock looked exactly as it had before, except for the position of the hands.

"Seven forty-five."

"Are you certain? Rotating a clock doesn't change the time, does it?" He had a point, but I didn't know what to make of it. "It only says four-thirty because someone decided that's what it means. What's on the wall is a dial with two hands, yet what you see is the time." I was still confused. He sighed, then continued. "All that's happened is that you've walked to the edge of the great mosaic of human knowledge. Up until now, you've been living in a world full of ideas and concepts that other people have set out for you. Now it's your turn. You get to design a piece of the mosaic and glue it down. It just has to fit with what else is there. And if you do a good job shaping your tile, it will be easier for the next person to fit his around yours."

"You're saying that I've been looking for an answer when really I should be making one up?"

He looked relieved. "Don't believe the bull about science being only an objective search for truth. It's not. Being a scientist also requires the skills of a politician. It's a struggle to define the terms, to guide the debate, and persuade others to see things your way. If you're the first one there" — again he pointed to the clock — "you get to say what it is that others will see."

As I drove back to my apartment, the answer to my problem came to me. When I got inside, I called Dr. Joshi and gave him a hasty review of my thinking. I could hear the sound of chalk against blackboard as he worked out the logic. After a long silence, he finally spoke. "Beautiful. Now all you have to do is write it up and get out of here. There's nothing else I can teach you."

Surely, I thought, he was being funny — this was just his way of complimenting me on a good idea. "Come on, that's not true at all!" I said.

"I suppose there *is* one other thing." He suddenly sounded more serious.

"What's that?"

"Just remember that ideas last longer than people or things. Your ideas will go further if you don't insist on going with them."

You know, he was not wrong.

1

The Idea

"Is this thing war surplus?"

"Huh?"

The taxi driver didn't get it. We were racing down a narrow road in the suburbs of Boston, lurching from pothole to pothole. Each bump rattled the vehicle as though a shell had exploded nearby. The maroon logo on the door read "Veterans Taxi." The driver was vintage antiwar sixties — short graying beard, ponytail held by a rubber band, and a Cossack hat with ear flaps as a concession to the bitter February cold. I was to meet Mitchell Kapor at Hanscom Field at nine A.M. sharp to check out his new toy, a personal jet. The trip from the Cambridge offices of Lotus Development Corporation — the company he had founded in 1982, only five years earlier — was supposed to take less than thirty minutes, but I was late, and lost. Mitchell had been clear that he wanted to depart promptly so we could arrive in San Francisco in time for his lunch appointment.

The pavement widened without warning, and a stoplight signaled our reentry into the civilized world. The access road circled the field to the Butler Aviation terminal, where the private planes were parked. As instructed, we drove through an unobtrusive gate onto the field. Several small planes and a single jet sat in the passenger loading area, randomly scattered like animals maintaining a safe distance at a communal watering hole. I was relieved to see Mitchell just ahead of us, pulling suitcases and tote bags from the trunk of his dark gray 1984 Audi sedan.

The unmarked jet was painted a nondescript brown and beige. A narrow gangway of four or five steep steps was carved out of its middle. Two large men in vaguely official dark blue outfits sporting epaulets and caps stood at ease on either side of the stairs, waiting for a limousine to deliver their new boss, the founder of the world's largest independent software company. They nervously eyed the two young men in blue jeans struggling toward them with bags hanging off both shoulders.

"Can we get some help, please?" Mitchell bellowed. The two men froze momentarily, realizing that this young guy with shirttails hanging out the back of his ski jacket was their man. They ran forward to relieve us of our luggage.

"Good morning, Mr. Kapor," one of the crewmen said.

"Call me Mitchell, and this is Jerry. He's hitching a ride today. We're splitting the gas."

Mitchell laughed at his own joke. The operating cost of the craft was more than a thousand dollars an hour, much of which was high-grade jet fuel. The crewmen glanced at each other in disbelief and then introduced themselves as the pilot and copilot.

We climbed the steps to find a cramped, tubular cabin decorated in dark brown fabric and wood paneling. It looked like a miniature old-fashioned men's club. There was a narrow aisle down the middle, just tall enough to stand in, with four seats along the right but only two seats along the left, followed by a couch long enough to lie down on. I imagined that the couch was there in case the jet's owner got lucky with a passenger — a sort of airborne version of the mattress in back of a pickup truck. Mitchell, a devoted family man, wouldn't see it this way, but I was single and more attuned to such possibilities. A custom-made bar, with cutouts for bottles, displayed the varieties of hard liquor favored by the previous owners — a bank whose executives had lived well before falling on harder times. There were also several Cuban cigars and packs of playing cards.

"We can get rid of this stuff," Mitchell said. "Some Diet Coke and sugarless gum would be fine."

His face impassive, the pilot made a note.

I first met Mitchell Kapor in 1984, when he wandered into my office unannounced and asked what artificial intelligence might mean to personal computers. I was a logical person to ask, having completed my Ph.D. in the field five years earlier.

After graduating from Penn in 1979, I joined the research staff of Stanford University. Stanford had the pace and style of a country club, with research grants blowing in through every open window. After slaving away for years on graduate studies and working every odd job I could find to support myself, I felt as if I had died and gone to heaven. It was a dream job, with virtually no responsibilities other than to think about something interesting and write up my ideas once in a while. In the absence of any objective measures of success, the tenured professors in the computer science department took to alternative means of establishing their self-worth, mainly by infighting and collecting academic titles. After about a year and a half of pastoral bliss, I concluded it was unhealthy to retire at the vital age of twenty-eight.

Early in 1981, everyone in sight was starting companies. I was unexpectedly offered the opportunity to join a new artificial intelligence company called Teknowledge, formed by a group of Stanford professors. Teknowledge built expert systems, computer programs that used knowledge gleaned from human experts to reason through complex problems, like diagnosing obscure forms of cancer.

Accustomed to the academic environment, the researchers did their work on large, symbolic computers called LISP machines, oblivious of the personal computer revolution taking place around them. The LISP machine was a classic boondoggle, built mainly under government grant and sold mainly to government research projects. An expensive, high-performance computer, the LISP machine was to the personal computer what an F-15 fighter jet was to a Cessna 150.

About two years into this endeavor, I suspected that similar results could be obtained at far lower cost on a personal computer. So I commandeered an IBM PC and started to write programs in my spare time. Within a few months, I had a number of promising prototypes up and running. In a remarkable coincidence, this was precisely when Mitchell came to visit, asking his question.

We immediately hit it off, and talked about how to design a flexible database to manage personal information — notes, ideas, to-do lists, phone messages, and the like — as opposed to corporate data such as billing and inventory records. Mitchell offered me a consulting contract to develop these ideas into a product, working directly with him and another scientist named Ed Belove. I could work at home, in the wooded hills just west of Stanford, with occasional visits to Lotus's offices in Cambridge.

For the next year or so, I lived and worked alone for extended periods, accompanied only by my cat, Critter P. Spats, the sole remaining evidence of a long-gone live-in girlfriend. Realizing that I might benefit from greater human contact, I took the proceeds from the sale of my Teknowledge stock and purchased a condominium on "crooked" Lombard Street in San Francisco. The constant flow of tourists down this cobblestoned landmark made me feel as if I had moved out of the wilderness onto the banks of the river of humanity. The cat loved the extra attention.

Shuttling to Boston about once a month, I worked closely on the Lotus project with Mitchell and Ed. Our efforts resulted in a new type of program we dubbed a personal information manager, or PIM. As the project neared completion, we officially named the product Lotus Agenda. In February 1987, I was hitching a ride back to San Francisco on Mitchell's new jet to show him some extra features we'd added to the product at the last minute.

Once we were on board, Mitchell started to search through his luggage. There were tote bags and briefcases everywhere. It was essential that the discriminating technophile travel with a variety of computers, portable phones, organizers, chargers, adapters, cords, and extra batteries, as well as the latest industry weeklies, computer magazines, and newspapers. I wondered if this was why Mitchell felt he needed his own jet — checking all this stuff on a commercial flight would be a nightmare. When he was comfortably ensconced in a fortress of electronics, he took off his ski jacket, revealing his trademark outfit: a formal Hawaiian shirt (white background) over loose-fitting jeans. Mitchell was a big man, nearly six feet tall, and walked with a boyish bounce. He had a wave of dark hair with a touch of gray at the temples, betraying his thirty-six years. His two front teeth were slightly askew, giving him the faintest aspect of a woodchuck, which was seconded by his zeal and diligence. I could see that he was perspiring lightly from our hurried boarding.

I looked like a junior Mitchell, the same height but twenty pounds lighter, though my hair was a bit more gray. The same ill-fitting designer jeans — crafted for some platonic *GQ* ideal, not a son of Abraham — curved under my waist and hung loose around my rear. Inevitably, the bottom button of my shirt fell above the belt buckle, leaving the shirttails free to wander their separate ways, revealing a roll of flesh. Like Mitchell, I was locked in perpetual battle with my weight, but the stakes were higher — I couldn't afford to carry the girth of a typical middle-aged husband, for fear of never becoming one.

We settled into the front pair of seats.

"Put your seat back in the full upright position, and fasten your seat belt tight and low across your lap," Mitchell admonished me with mock seriousness.

We spent the next several minutes repeating verbatim the inescapable Big Brother rituals of the commercial airlines. We were soon laughing hysterically, and the pilots must have thought we were nuts. After taxiing a short distance, we were off the ground, climbing at a steep angle. We sat in silence for the next few minutes, watching the ground recede and feeling very regal.

As soon as we leveled off, Mitchell pulled out his latest gadget — the lightest, most powerful portable computer available. This remarkable machine, the Compaq 286, packed all the power of the most recent generation of desktop personal computers into a box about the size and weight of a small sewing machine. The numeric designation 286 was not selected at random. It indicated that the product contained at its core a microprocessor chip called the 80286, designed and manufactured by Intel Corporation.

In the mid-1980s, computer cognoscenti had a penchant for substituting technobabble for plain talk. This served a useful purpose. Learning to use a computer — much less to program one — required a level of personal commitment commensurate with learning the piano, and a similar level of innate talent. It attracted people who had difficulty with the messy business of human relations, preferring instead the company of predictable and infinitely patient machines. This devotion was rewarded with valuable skills and friendships. Former wallflowers suddenly found themselves accepted into a new society of like-minded people who were more comfortable communicating through electronic mail than face to face. Now they could mask their awkwardness behind CPUs, RAMs, and modems. Geeks became chic.

A secret language was the key to the club, like the lingo used by each generation of teenagers to identify kindred souls and exclude ignorant grownups. It made the members of this new caste feel special, smarter than everyone else. The embarrassment that ordinary people felt about their lack of computer knowledge only reinforced this feeling. But just knowing the model number of a computer wouldn't help you join the secret society — you had to know how to *pronounce* it. Nowhere was it written that 80286 should be read "eighty, two eighty-six," as opposed to "eight-zero-two-eight-six" or some other variation. Welcome to the club.

"Bear with me," Mitchell said. "I've got to update my notes." He began to rifle through his pockets, pulling out small pieces of paper. Some were yellow stickies, some were pages ripped from a spiral-bound pocket notebook; there was even the stray napkin or gum wrapper. Mitchell was a prolific note taker, jotting down every interesting idea and reference within earshot. You never knew when he was going to strike out and appropriate the nearest writing implement and fragment of paper. On one particularly frantic occasion, I saw him tear the corner off a page of the *New York Times* and write in the L-shaped margin.

Making sense of this motley collection of ideas, phone numbers, and reminders was Mitchell's passion. That's why he was so committed to Lotus Agenda — he desperately needed the product himself. He powered up the Compaq 286 and waited while the machine went through its lengthy startup process.

Mitchell rolled his eyes, whistled, and tapped his foot in exaggerated impatience. With Agenda finally up and running, he began typing in his accumulated scraps of notes with the efficiency of an executive secretary.

"You know, this is really the pits," he said. "It doubles the time it takes for me to keep organized. I wish there was some way for me to get all this stuff directly into the computer and skip the paper."

That sounded like a challenge to me.

"Look, Mitchell, it seems that the real question is how small and light you can make a portable computer."

"Well, what are the largest components?"

I thought for a second. "The disk drives are one problem. They probably weigh about two pounds each. Next is the power supply, the battery. Another few pounds. The display isn't really that much, but the two layers of glass are pretty heavy. Next is the keyboard, I guess. It doesn't weigh a lot, but it's pretty big."

"Yeah, and you can't shrink it much and still be able to type normally," Mitchell added. "Below a certain size, you'd have to poke at it with a finger rather than really type. That's why the pocket organizers stink. They're really just overgrown calculators that you can stick a few names and phone numbers in." He was referring to gadgets like the Sharp Wizard that let you store a small amount of information in addition to doing arithmetic. These products were purchased on impulse and, inevitably, soon abandoned in a drawer until the batteries leaked.

"Other than the keyboard, though, it ought to be possible to build something pretty cool, sort of flat like a book," I said. "You could have a flat display with all the electronics on a single board just behind it, and maybe a flat battery, like a sandwich. Instead of disks, you could use SRAM for the data." SRAM stands for static random access memory. It is like normal computer memory but requires much less power to keep the memory "alive," and so places only a modest demand on a battery to retain its contents while the computer is turned off.

"The interesting thing is, in principle you wouldn't need to load and save files. You could just leave the programs running and the files loaded all the time." Mitchell was getting into it now. In fact, we were both cooking. This was the fun part of the business, imagining possibilities. With technical breakthroughs announced every week — smaller disks that stored more data, faster microprocessors, denser memory chips — anyone could combine these parts and build a new product or create a new market. "But the problem is still the keyboard," he said.

"Not only that, it's not always practical to type when you want to take a note." I waved my hands in the air as though typing on an imaginary keyboard. "You have to set the computer down somewhere to free up your hands."

Mitchell picked up on my thought. "Even in meetings it's a

problem. Only propeller heads use laptops in meetings, and it's always distracting."

We continued exploring this line of discussion for some time until the pilot stuck his head into the cabin, bearing a plastic tray of half sandwiches, some cut-up vegetables, and a container of dip. "Waddaya want to drink?" he asked.

Mitchell and I looked at each other, realizing that we weren't going to get the usual choice of beef, chicken, or "today's light fare." After pulling off the plastic wrap, we picked our way through the soggy assortment of mystery meats, topping off the meal with a delightful medley of miniature candy bars.

After lunch I settled in for a brief nap, tired out by Mitchell's intensity and the carbohydrates. Besides, I got my best ideas in my sleep.

Mitchell Kapor had wandered into the computer business along with lots of other underemployed but energetic young men at the time. The difference was that he had struck gold, while most others were still panning dirt from barren streams.

In the late sixties, times were tough for college kids with a social conscience — the deaths of Bobby Kennedy and Martin Luther King, the My Lai massacre — which lured Mitchell away from starting down the Wall Street fast track and toward a life of manning protest barricades and experimenting with psychedelics. Like many, he eventually concluded that writing software for personal computers seemed a promising path. He learned the BASIC programming language and ultimately landed a contract to develop a business graphics package for Personal Software Corporation, later known as VisiCorp, which produced the hugely successful Visicalc spreadsheet program.

His product, Visiplot/Visitrend, read data from the files saved in Visicalc and displayed the numbers visually on graphs and charts. While this was a handy thing to do, it was awkward to save a file, exit one program, start up another, and load the file again just to get a look at a graph. So Mitchell conceived of a new and better product that integrated these functions. As required by his contract, he offered the idea first to Personal Software. The company passed.

By the fall of 1981, a special breed of investor was scouring the

Silicon Valley like a big league scout, searching for young entrepreneurs with promising ideas. These venture capitalists — usually called VCs — were known to put a million dollars on the table overnight for the right opportunity. It was a modern California gold rush.

This overheated climate created a feeding frenzy around personal computers and software. Mitchell approached Ben Rosen, a former Wall Street analyst, and L. J. Sevin, a colorful Texan, with his idea. This unlikely pair had formed a partnership to grab a piece of the action, which they did with a vengeance. Rosen had used some of Mitchell's software, and liked it. They cut him a check and went on to their next deal. Mitchell named his new company Lotus Development Corporation because of its countercultural, oxymoronic tone, like the name Apple Computer.

Lotus was in a race with several more established competitors, and he knew it. Most industry people thought Lotus was crazy to try to unseat Visicalc, which dominated the market at that time. But Mitchell intuitively understood something that only the Japanese taught in school: in a fast-growing market, what matters is your share of *new* machines, not existing ones, because new machines quickly come to dominate the market. When IBM announced that it would be bringing out its own version of the personal computer, to be called the IBM PC, Mitchell directed his programmers to tailor their program to this machine.

In 1982 a crop of new and improved spreadsheet products were slated to debut at COMDEX, the computer dealers' exposition held every November in Las Vegas. The Lotus team scrambled to ready their product, which they named 1-2-3. Mitchell decided that to beat the competition, Lotus had to come out of the starting gate with a bang, so he committed the unheard-of sum of $1 million to marketing and promotion. Mitchell knew it would all be over by Christmas, one way or the other — they would ride the IBM PC into orbit or be marooned in the Las Vegas sand.

As the last dusty moving vans carted away the exhibit booths from the Las Vegas Convention Center, it was clear that Lotus 1-2-3 was a major winner. Lotus's heavy promotion had created a larger-than-life impression, and more important, one that dwarfed its competitors.

Rosen and Sevin knew they had hit the jackpot, and soon trotted the fatted calf off to Wall Street for sale in an initial public offering. With the IPO, Lotus became the darling of investors and an instant emblem of the success of the American capitalist system. Mitchell Kapor, still a compassionate and philosophical proponent of the counterculture, found himself invited to meet the governor of Massachusetts and was profiled in the *New York Times.* It was as though the megaphones had abruptly gone silent and someone had politely explained to him that it was all some terrible mistake — would he please come across the police line and tell the nice officials what he thought they ought to do. To top things off, he was now rich.

Half an hour later, I woke up with the strange feeling that I had just forgotten something important. I turned to Mitchell, who was still clicking away on his Compaq 286, and stared at him.

"You OK?" he asked.

I was quiet for a moment, trying to get my bearings. Suddenly I knew what was bothering me. "Mitchell, suppose you used a pen instead of a keyboard."

"What do you mean?"

"Suppose that instead of typing in text, you write with some sort of stylus directly on the screen. If it was possible to sense where the tip of the pen was, electronic ink could appear on the screen right under it, and it would look like you were drawing a line, or writing. The trick is to turn the handwriting into regular text or numbers. I'm sure there are algorithms that do this."

"What about the function keys?" Mitchell was referring to the non-alphabetic keys on a keyboard that allow you to issue commands like paging up or down, saving or loading a file, deleting a word, and so on.

"I don't know. I guess you could figure out some way to do that stuff with the pen too — like tapping on icons or menus. The point is to eliminate the keyboard."

"A device like that would be more like a notebook or pad of paper than a laptop," Mitchell observed. "In fact, the thing would be so different, you'd need a whole new approach to software."

We both sat in stunned silence as this insight sunk in. People had certainly thought about pens and computers before. We had both seen electronic tablets attached to desktop machines, which were mainly used by drafters and artists. But these tablets had always been just another type of input device, an accessory to the keyboard and the mouse. This idea was different. We had put it all together, combining simple, familiar elements into something radically new. We recognized that an electronic pen, without a keyboard, could create a completely different device.

Almost like clockwork, every ten years since the beginning of the computer revolution, a new class of computers had unexpectedly emerged.

In the 1960s, the popular image of a computer was a room full of mysterious, monolithic boxes the size of refrigerators, with spinning tapes and banks of flashing lights. These remote, intimidating mainframes were housed in special rooms with raised floors, filtered air, and glass walls which looked like an intensive care unit for electronics. With the advent of the minicomputer in the 1970s, the popular conception of a computer began to shift. These were smaller boxes that could be hidden in a closet or storeroom and connected to terminals sitting on the desks of engineers, bank tellers, and travel agents. In the 1980s, the personal computer took its place as the dominant form of computing. Now a computer was an individual productivity tool, consisting of a video screen sitting atop a rectangular box with a slot for floppy disks.

This historical progression was not just about the physical look of the machines. With each wave of devices came whole communities of customers, operating systems, applications software, and companies. Subsequent generations didn't supplant the previous ones; they empowered different people by solving different problems. Like some profound new form of energy, the underlying constant — computing power — had no shape or substance. Each type of computer simply delivered this power in a different way.

Until that moment on Mitchell's jet, it had been difficult to imagine what the next generation of computers might be like. There was an obvious trend toward faster, smaller machines, but it was hard to see how the basic character of the computer would change.

Entrepreneurs had made so much money from the personal computer industry that many believed the computer revolution had achieved a kind of final, perfect form. It seemed that the personal computer was right for all people — they just didn't know it yet. It was only a matter of time before everyone, from the CEO to the youngest schoolchild, would use one.

Mitchell and I had stumbled upon a plain truth: personal computers are deskbound, like typewriters, and are unsuitable for people who spend their time away from a desk or work face to face with others. There are segments of the population that fit this profile — salespeople, inspectors, consultants, and delivery people, to name a few. They certainly have access to computers back in their offices, but this isn't where they really do their business. Moreover, the problem isn't just taking a computer with you, it's how you use it in the course of your work. In meetings with colleagues and clients, mobile professionals use pen and paper — notebooks, folders, and calendars — because these don't interfere with the interpersonal communication so critical to their jobs. Still others — the clipboard crowd — need to write while they are standing up or walking around.

Mitchell and I both instantly understood that the key to the next wave of computing was to create a device that worked like a notebook instead of a typewriter. The impact of this insight, so obvious and yet so novel, overwhelmed us. We shared a sense of seeing something utterly new and important, like the adventurers who had first encountered the Rocky Mountains, now five miles beneath our feet.

This unique emotion — the modern scientific version of religious epiphany — is startling in its raw power and purity. I had experienced it previously only once or twice in my life, always at an unexpected moment after months of struggling with an impossible problem. It is reserved for those rare occasions when you, and you alone, know something that no one else knows. The poet Edna St. Vincent Millay wrote that "Euclid alone has looked on beauty bare." And now Mitchell and I knew exactly what she meant.

We were momentarily unable to speak. I saw Mitchell's eyes become glazed and teary.

As we started our descent into San Francisco, we regained our composure and began fleshing out the details of the concept. The tone had abruptly shifted from an intellectual exploration to something very personal. We knew we had to pursue this. "You do realize what this means." Mitchell's voice tightened as he spoke. "A whole new operating system, a whole new model of how the user will work with the system."

"Hell yes," I said. "There's a lot of far-out research to be done. The first thing is to answer some basic questions, like what resolution is required to display writing legibly on a computer screen, and whether it's possible to write directly onto the surface of a flat display."

"Then, of course, it all hinges on how good the handwriting-recognition software is." We both said something like this simultaneously, and laughed as though we had bumped shoulders trying to walk through a door at the same time.

"Hey, look," Mitchell said, "I'm having lunch with this great guy at Apple, Steve Sakoman, who's sick of the place and wants to leave. He's in charge of hardware engineering for the Macintosh, but he wants to do something more exciting with portable computers and he isn't getting any management support. I'll talk to him about this and see if he thinks it's feasible."

"Since this most likely involves a new operating system, maybe Gates would be interested." I was referring to Bill Gates, the techie founder and chairman of Microsoft, at that time the second-largest PC software company after Lotus. Mitchell stared at me with a dark intensity.

"You don't understand. The PC business is *war*." He spoke as if announcing this to a crowd. "Either you fight or you're a casualty. You have to look the enemy in the eye and never, ever blink."

I wasn't sure why Sakoman at Apple was an ally and Gates at Microsoft was an enemy. But clearly that's what Mitchell meant.

As the plane taxied to the private aviation terminal at San Francisco Airport, Mitchell and I packed up our gear, ready for a quick getaway. His rental car was prepped and waiting next to the plane as we rolled to a stop. A man in a Hertz shirt snapped to attention, ready to collect Mitchell's signature and hand him the keys. For the big brass, the rental companies don't mess around.

The gangway sprung open with such force it rocked the plane. "I'll call you tonight," Mitchell said, and took off south to Cupertino like a guided missile. As I hiked to the parking lot to pick up my car, I could tell I wasn't going to get much sleep that night. Or, for that matter, for the next few years.

⊡

Within a couple of weeks we had become thoroughly enchanted with the idea of a pen computer. Over the previous few years Mitchell had grown bored at Lotus, where he was increasingly called upon to play a ceremonial role, presiding at product announcements and board meetings. As he shed his operating responsibilities, in part to work on Agenda, he joked that he was becoming a rabbi — reduced to reciting blessings on special occasions. The Agenda project had rekindled in him the unique feeling of working with a small, close-knit team of engineers, reminding him of the early Lotus days. But Agenda was just a taste, and he wanted more. Perhaps the idea of a pen computer was a way to get it — if he could build the right kind of spirit.

After several preparatory one-on-one meetings, he called together his hand-picked team of collaborators for the new project. In addition to Mitchell and myself, it included Peter Miller, a brilliant and erudite software project manager from Lotus, and Steve Sakoman, the restrained and thoughtful hardware engineer whom he had visited at Apple Computer in Cupertino. We met in a suite at the Kendall Square Marriott in Cambridge. Outside, the last piles of salty winter slush were fading in the fresh spring air.

Peter was like a walking *New York Times* crossword puzzle — an elaborate tessellation of obscure facts fitted together into a pleasing pattern. His pot belly and halo of gray hair made him look like a latter-day Friar Tuck. In contrast, Steve had a clinical, focused style. He dressed in various shades of plain, like a Norman Rockwell painting of a practicing engineer. His nondescript chinos precisely fitted his lean frame; all pocket flaps were neatly pressed flat and buttoned. Mitchell opened the meeting by unrolling several archi-tect's drawings on the conference table.

"As you know, I've taken a lease on half a floor of office space in the tower next door. Here's my thinking about how we could lay out the offices . . ."

For anyone else, this would have been a classic "presumptive close" — proceeding as though an agreement had already been struck among the participants. In Mitchell's case it was raw enthusiasm harnessed for a special purpose: it allowed the team to practice collaborating on a non-threatening subject. We studied the documents. To my surprise, a spirited discussion ensued.

When Mitchell had achieved an acceptable level of give and take between the participants, he steered the discussion to the core issues: equity, financing, and the nature of the project itself. It slowly emerged that each of us had a different agenda. He was interested in building a new company that would recapture the teamwork and momentum of the old Lotus without repeating the same mistakes. Peter was interested in designing an "object-oriented" computer interface that would knit together diverse documents, spreadsheets, and databases in a coherent, organic whole. Steve wanted to build the leanest, flattest computer possible. I wanted to deliver a practical pen computer for working professionals. In each case, our goals for the project reflected our respective personalities and visions of a better world.

The meeting stretched on until late in the afternoon, by which time Mitchell had deftly built a strong sense of unity, despite the fact that the project was still far too vague and broad to be practical. Each of us took our action items: Mitchell would line up financing, legal counsel, and facilities; I would investigate handwriting recognition; Peter would think about object-oriented interfaces; and Steve would notify his superiors at Apple that he would be leaving to join our venture.

We finished up with an easy topic: naming the new company. We wanted something short and striking that would project a sense of mobility. Mitchell thought we should consider a two-letter word as a name, and proposed GO. I suggested ON. Peter observed that "On Technology" contained the word "ontology," the study of being. So we went with ON.

▣

Forming a new company is like starting a romantic relationship. The early phase is emotionally volatile. You have to build confidence, establish a sense of fairness and balance. If one person feels he is investing more of his feelings, without reciprocation, the situation can quickly get out of hand, resulting in stormy mood swings ultimately leading to disaster. Continual contact and reassurance are essential. That's why the three-week gap until our second meeting was a dangerous interval.

The meeting was to last for two days, over a weekend. I was staying at Mitchell's house, and Steve stayed at a hotel. At the outset Mitchell worked hard to recapture the enthusiasm of the earlier meeting. However, I noticed that Steve seemed distracted and less involved. That evening, after our first session, I mentioned my concern to Mitchell, who hadn't openly acknowledged Steve's change of mood. Mitchell restrained his natural exuberance and sat quietly for a moment.

"What do you think is happening?" he said.

"Maybe he's having second thoughts. He's being asked to give up a lot — a high salary, great benefits, a secure senior position at Apple, moving his family to Boston —"

"Or maybe he went in to quit, told them what we were going to do, and they made him a counteroffer," Mitchell suggested.

I was a bit startled by his sudden paranoia, but then realized that this was distinctly possible. Steve was a free agent, and at this stage no one really owned the ideas we had been batting around. He was well within his rights to pursue the project at Apple if he could get the required internal support for it.

The mere thought that Steve might defect caused Mitchell to feel betrayed. It undermined the sense of teamwork, of Musketeer-like commitment, that he was seeking to create. Even worse, Steve was in a position to walk off with the project and put the power and resources of Apple behind him.

Throughout the night, Mitchell's mood grew darker. Unable to check his suspicions out with Steve until the following morning, he became increasingly fearful of losing the fragile flame he had worked so hard to spark. By morning he was quite agitated. At his first opportunity, he confronted Steve.

"Steve, I'm a little concerned that you may have some doubts about proceeding with the project."

Always honest but tempered, Steve came clean. "Well, I talked to Jean-Louis, and he escalated the matter to Sculley." Jean-Louis Gassée, an opinionated and astute Frenchman, was president of Apple products, and he held John Sculley — the non-technical CEO who succeéded Steve Jobs — under a Svengali-like influence. Apple had a long-standing tendency to canonize its technical gurus, which was probably exacerbated by Sculley's personal doubts about his suitability to lead a technology company.

"I met with Sculley, and he asked me what it would take to stay and do the project at Apple," Steve said. "I gave him a straight answer — complete freedom, protected resources, a separate staff and site. He agreed to my terms. I have to give him a final answer this week."

Mitchell's worst nightmare had just come alive. But he had had an entire night to prepare an emotional defense. He spoke to the team. "You know, guys, I love this idea, but I think maybe we're moving too fast. Maybe I'm not ready to jump back into startup-land right after Lotus. I'd like to take some time to consider what's really going to work for me and what's not." Under the circumstances, this was a measured response, but I believe it was driven as much by personal disappointment as self-analysis.

That was it. Just as easily as the idea had started, the project was dead. "Easy come, easy go," I whispered to Peter as Mitchell and Steve attempted to make polite conversation.

But after the meeting broke up, Mitchell took me aside. "Jerry, I've been thinking. Why don't you try to do this project yourself? You're a smart guy . . ."

I was shocked. I thought of myself as an engineer, not a CEO. Mitchell had a reputation as a deft manager and business strategist. "But I've never managed squat."

"You think I had any more experience when I founded Lotus? Just remember, political history is written by whoever wins the war, and corporate history is written by successful entrepreneurs. All you have to do is make a pile of money and everyone will think you're a genius." He chuckled and shrugged. "C'mon, I'll introduce you to some VCs."

2

The Deal

THE MODERN CORPORATION is quite possibly the highest form of human cooperation. Specialized resources in the form of labor, raw and finished materials, capital, and knowledge come together in a marvelous process that transforms these components into goods and services of greater value. This miraculous conversion is similar to the process by which dirt, water, sunlight, and a packet of information in the form of a seed are reorganized into a living plant. Like plants, corporations are born, grow, and die, reach out for resources, fend off predators, and compete with others. Businesses evolve over time, as less efficient corporations are replaced by more effective ones, whose successful practices are then emulated by others.

It is against this backdrop that the "startup game" developed. The startup game is an elaborate contest created to accelerate the pace at which corporations evolve, played continuously by an endless parade of hopeful entrepreneurs. It is a carnival game of life, testing the strength, aim, and skill of contestants willing to expose themselves, for glory or ridicule, to public scrutiny. The challenge is to find a new or better way to do business; the rewards are increased wealth, enhanced personal reputation, and control over one's own destiny. The startup game is designed to motivate our brightest, most creative, and hardest-working individuals to improve the use of society's resources, increase employment, and provide a broader range of quality goods and services. Here's how the game is played.

It begins with an aspiring entrepreneur who is willing to step right up and be tested. As in many other games, the player starts with an artificial currency — in this case, the stock of the new venture. The goal is simple: increase the value of the entrepreneur's shares, because when the game is over, these can be cashed in for real money. The trick is to swap some of the stock for three resources — ideas, money, and people — then use these resources to increase the value of the remaining stock.

The ideas are called intellectual property, which includes the business concept itself and any unique designs, processes, or plans for how to pursue the business. The intellectual property is safeguarded by trade-secret protection programs and by patents and copyrights.

The money comes from investors, usually a venture capital partnership. The venture capitalists, or VCs, get a special form of equity called preferred stock. This stock confers certain rights, known as preferences, such as the right to appoint someone to the board of directors.

People are usually assembled by recruiting friends and associates and by hiring headhunters to find them. These people become the employees, who are paid with a mixture of cash and stock. The cash is enough to cover reasonable living expenses, but is often less than the employees would earn at a stable, going concern. The stock is what they are really working for. It is a part ownership in the venture, a chance to participate in the game and pay for it with their own labor — which is why it is often called sweat equity.

The entrepreneur doesn't raise all the required money up front because that would mean selling too much of the stock. Usually, the initial investment is just enough to reach some identifiable milestone. This milestone is chosen to demonstrate to potential future investors that the company's prospects have improved, justifying a higher price for the stock, so that less of it has to be sold. If the money runs out before the milestone is reached, the game is over. In the meantime, other companies may try to steal the ideas, the people, and may even try to run the venture out of money.

Venture capitalists typically look for a reasonable possibility of making five to ten times their original investment within five years. This corresponds to borrowing money at 50 percent per year or more (until recently, such an interest rate would have been considered criminally usurious). This high cost of capital makes the startup game a race against the clock.

The final step in the game is a financing event called an initial public offering, or IPO. The IPO — when the company is first listed on a public stock exchange — usually marks the transition of the enterprise from a risky venture to a profitable company. Up until this time, it is virtually impossible to dispose of stock and get cash. Soon after the IPO, the entrepreneur is free to sell stock on the open market, as are the investors and employees. This marks the successful end of the startup game and the creation of a viable company. The entrepreneur can now cash in his stock and go home, take on the less risky job of managing a going concern, or step up and play the game again.

◻

Soon after the disastrous meeting in Boston, Mitchell arranged an introduction to the legendary venture capitalist John Doerr, whose firm of Kleiner Perkins Caufield & Byers had backed such prominent technology companies as Tandem, Compaq, Sun, and Lotus. John asked me to drop by his office the following Monday to chat with some of his partners. I mistook this for an informal bull session, but any experienced entrepreneur would have known better. To VCs, Monday is as sacred as Sunday is to the Vatican. This is the day when venture capital partnerships around the world have their official meetings to review potential investments. All the partners participate, come hell or high water — by video conference call from a branch office if possible, or by speaker phone from their deathbed if necessary. An invitation to address the partners at a top-tier firm like Kleiner Perkins — or KP, as it is known in the trade — is an unusual opportunity. I was unprepared.

At the appointed hour, I showed up wearing a sport jacket with my shirt open at the collar. I carried little else but a maroon leather portfolio holding a tablet of paper and a pen which I had received as a Christmas gift. No business plan, no 35-millimeter slides, no charts, no financial projections, no prototypes.

The Kleiner Perkins offices were on the thirty-fifth floor of a posh office tower in the heart of San Francisco's financial district.

The floor-to-ceiling windows framed a spectacular panoramic view of the bay on one side and the city on the other. The partners' offices were extravagant by most standards, set apart from each other and from a large common area by smoked-glass partitions.

The previous presenter was just finishing up. Dressed in a dark blue pinstripe suit, his red power tie thrust forward with a gold tie pin, he was nervously fielding rapid-fire questions from a brigade of partners and associates inexplicably packed into a small corner conference room. A bare prototype circuit board sat on the conference table. A crisp color graph projected on a whiteboard; the man's perspiration gleamed in the reflected blue light.

"Thank you, we'll let you know our decision in about a week," said one of the partners. The presenter collected his belongings and left quietly.

After a short break, John Doerr reassembled the group and invited me in. He made a short introduction, covering my background and explaining Mitchell Kapor's interest in the project, then turned the meeting over to me. I was nearing a state of panic. I paused to size things up, knowing that the brief silence before diving in would create a momentary impression of authority. In reality, I was searching for a strategy.

I had a flash of déjà-vu. I remembered facing this same kind of "show me" crowd at my Ph.D. dissertation defense. The key to success there was in recognizing that although the examination committee had the power, I had the knowledge. The same was true here. I had done a lot more thinking about this topic than anyone else in the room, and to win their respect, I merely had to demonstrate this fact by keeping the discussion focused on the areas I had already investigated. I decided to lead with the business issues.

"Gentlemen, you probably think that there is no longer any way to make money by starting a new personal computer company. The competition is brutal, and the barriers to entry are high. However, I'm here to suggest to you the possibility that the PC as we know and love it may not be the best and final form that computers will take. I believe that a new type of computer, more like a notebook than a typewriter, and operated by a pen rather than a keyboard, will serve the needs of professionals like ourselves when we are away from our desks. We will use them to take notes; send and receive messages through cellular telephone links; look up addresses, phone numbers, price lists, and inventories; do spreadsheet calculations; and fill out order forms. All of this can be done unobtrusively while sitting in meetings, conferring with clients, commuting to work on the train, or even when standing up and walking around. Like the fax machine, the pen computer can dramatically accelerate the pace and increase the efficiency at which business can be conducted.

"I can't say for sure when this will occur. But I do know that it will happen, and someone is going to make a big pile of money on it. With a little luck and some hard work, I think we could be the ones. Like the PC, I think this concept could come out of nowhere and take the industry by storm. I want to be the one to do it."

Having covered the business angle, and thrown in a gratuitous pledge of personal commitment, I proceeded to talk about the required technologies and the state of the art. The plain fact was that the project was technically very risky, and there was no point in hiding it. The greatest risk was whether a machine could reliably recognize handwriting and convert it into its equivalent in computer text, known by the strange acronym ASCII.

My audience seemed tense. I couldn't tell whether they were annoyed by my lack of preparation or merely concentrating on what I was saying. Several people narrowed their eyes disapprovingly — or perhaps they were just deep in thought. I had been talking nonstop for about ten minutes, and figured I'd better close. Thinking I had already blown it, and therefore had little to lose, I decided to risk some theatrics.

"If I were carrying a portable PC right now, you would sure as hell know it. You probably didn't realize that I am holding a model of the future of computing right here in my hand."

I tossed my maroon leather case in the air. It sailed to the center of the table, where it landed with a loud clap.

"Gentlemen, here is a model of the next step in the computer revolution."

For a moment, I thought this final act of drama might get me thrown out of the room. They were sitting in stunned silence, staring at my plain leather folder — which lay motionless on the table — as though it were suddenly going to come to life. Brook Byers, the youthful-looking but long-time partner in the firm, slowly reached out and touched the portfolio as if it were some sort of talisman. He asked the first question.

"Just how much information could you store in something like this?"

John Doerr answered before I could respond. "It doesn't matter. Memory chips are getting smaller and cheaper each year, and the capacity will probably double for the same size and price annually."

Someone else chimed in. "But bear in mind, John, that unless you translate the handwriting into ASCII, it's likely to take up a lot more room." The speaker was Vinod Khosla, a young man born in India and educated at the Stanford Business School, who was the founding CEO of Sun Microsystems. He acted as a consultant, helping the partnership evaluate technology deals. His cool manner masked a fierce analytical mind and the competitive instincts of a gladiator.

Before John could respond, Frank Caufield, a partner known for his outspoken manner, pointed at him. "Storage won't matter with *John's* handwriting — it's completely illegible!" This was the break I needed.

"Then you won't mind if I use his writing as a benchmark of our progress," I shot back. The rest of the group fell silent and looked toward Doerr and Byers, to gauge their reaction to my impertinence.

They both chuckled, signaling their approval. Then the entire room exploded with laughter. "At least I know how to type," John said.

From that point on, I hardly had to speak. The partners and associates traded good-natured barbs and insights as we fleshed out this new business opportunity. The single conversation split into two threads, then split again, as everyone offered a question or an opinion. It was beginning to sound like a lively cocktail party. Periodically someone reached out to touch or examine my portfolio, which was bouncing back and forth among the participants. It had been magically transformed from a stationery-store accessory into a symbol of the future of technology.

This din continued for several more minutes, after which Brook Byers called the meeting back to order.

"One last question," Byers said. "What are your personal goals for this project?"

I spoke slowly. "Well, I'm not in it to prove that I'm smart, or to prove that I can run a big company." I knew these to be two of the most common mistakes that would-be entrepreneurs make in starting a company. "I guess I have four goals. First, to produce a product that delivers real value to real people. Second, to provide an above-average rate of return for the investors. Third, to create a healthy and challenging work environment for the employees." I paused.

"And fourth?" Byers said, raising his eyebrows.

I couldn't think of a final point. I glanced around and noticed a plate of sandwiches and cookies left over from their working lunch. "To never pay for a meal for four years."

He laughed. "Have some. We're a full-service venture capital firm."

"Thanks." I picked up a cookie.

Then they took a break. Several people scrambled for control of the conference room telephone. The losers bolted out the door in search of another line. John Doerr turned to me and said, "Good job. I'll talk this over with the partners and let you know where we stand."

I thanked him, recovered my portfolio from two associates who were inspecting it closely, and left.

On my next visit to Cambridge, to do some final consulting work on Lotus Agenda, I was abruptly awakened by the phone in my hotel room. It was John Doerr. How he tracked me down I don't know, but it was all the more strange because of the time difference — he was calling from San Francisco at four-thirty in the morning. Before I could get my bearings, John got down to business. "On behalf of the partnership, I'm pleased to tell you that we've decided to back you."

Now I was totally confused. I hadn't given him a business plan yet, nor asked him directly for any money. "John, I'm very happy to hear it, but don't you want to see some financial projections?"

"We're backing you and the idea." That seemed pretty straightforward, but at that hour of the morning, I'm never quite sure of things no matter how obvious they seem.

"We should get together right away to close the deal," John said. "When will you be back?"

"I'll be leaving for San Francisco tonight."

"Too bad, that's a problem. I'm leaving for New York first thing in the morning."

"When will you be returning to San Francisco?" I asked.

"Not for a week, and that's too long."

I didn't know John well at that point. He was a slender man with thin, sandy hair and a deep voice. He often wore khaki pants and a white button-down shirt, and always lugged around several canvas bags full of papers and electronic gadgets. When excited about an idea — which was most of the time — he would jump up and pace around like a whippet in a cage. I had never before known anyone so passionately committed to devoting every ounce of energy to his life's work. Once he had set a course of action, he pursued it with the intensity of a laser and the impatience of a speed freak. His style was to cut through problems with dramatic solutions of grand proportions, as I was about to discover.

"What airline are you on?" he asked.

"TWA."

"OK, they have a hub in St. Louis. I've got an 'OAG' here, just a minute." He was referring to the pocket-sized "Official Airline Guide," a standard accessory for executives with a heavy travel schedule. "Change your flight to the nine A.M. tomorrow from Logan. That will arrive in St. Louis around ten fifty-five. There's a flight to San Francisco at twelve. That will give us an hour. I'll arrange a connection through there as well, and meet you at the gate."

"Will do," I said.

The next morning, I was just starting to perk up as the flight pulled into the gate in St. Louis. As promised, John was there to meet me. We soon agreed on a deal: the backers would get 33 percent of the company in Series A preferred stock in return for $1.5 million, which would be split among Kleiner Perkins, Mitchell Kapor, and a few smaller individual investors including Vinod Khosla, who had taken a keen interest in the project. John would be chairman of the board, I would be CEO. The board of directors would consist of John, Mitchell, Vinod, and me. Of the remaining equity, I would personally get 25 percent, and two other people — a VP of software and a VP of hardware, who were not yet identified — would each get one third of that amount (8.3 percent). The rest would be reserved for future employees.

With the business settled, we headed back to the gate for my flight.

"What do you want to call the company?" John asked.

I thought for a moment. ON was taken — Mitchell and Peter Miller were talking about some other ideas and were likely to use this name for whatever they decided to do. My second choice had been GO.

"GO, all caps," I said. "As in GO forth, GO for it, GO for the gold."

"As in GO public," John added.

When I boarded the noon flight, I realized that when the deal closed, the idea was worth $3 million (the total valuation of the company minus the invested cash), more than $1 million of which was mine — if I could ever convert that value into cash. And, of course, the investors were counting on its being worth a lot more.

◻

Having assembled the idea and the money, the missing resource was people. I knew two engineers who I thought might be suitable and interested. One was Robert Carr, the bookish, articulate chief scientist of Ashton-Tate, which was then the third-largest PC software company (after Lotus and Microsoft). The other person was Kevin Doren, a brilliant and practical hardware designer whom I had worked with several years earlier, on a project to develop the first all-digital keyboard music synthesizer, before he founded his own company in Miami.

I called Robert and made an appointment to meet him for lunch at a small bistro in the western part of San Francisco. I had met him only once before, and didn't quite remember what he looked like: young, a bit short and muscular, with brown hair and eyes. He resembled an earnest high school honor student who went out for middleweight wrestling. What I did remember was how hard he was to read. You could never tell what he was thinking by looking at him.

We picked a table in the back of an outdoor courtyard, surrounded by concrete and shrubbery. As soon as we ordered, the coastal fog began to roll in and the temperature dropped precipitously, providing us some privacy when the damp chill drove most of the other diners indoors. We buttoned up our jackets and I explained the concept once again. Repeating the act that had worked at Kleiner Perkins, I brought the same leather portfolio and used it as a prop. Robert was strangely quiet during my presentation, mostly looking down at the case. His brown eyes stared through his round wire-frame glasses as I explained the technical challenges and the business opportunity. As I finished up my pitch, I was consumed by curiosity.

"Any questions?"

He sat there silently, staring at my portfolio.

I waited, becoming impatient with his unresponsiveness. Finally I asked, "So, what do you think? Any chance you might want to do this?"

He slowly looked up, and I saw that his eyes were watery. I had mistaken his silence for lack of interest, when in fact he was struggling to keep his feelings in check. He was experiencing the same epiphany that had overcome Mitchell and me on that first flight.

Then he spoke. "Jerry, it's not a question of *whether* I want to do this. I *have* to do this. This is important. This is profound." He picked up the portfolio and lofted it up and down for emphasis. "You know, I'm happy at Ashton-Tate, but it's not very often that opportunities like this come along — something really big, a chance to really make a difference. Maybe once a decade or so. I think you've got one here."

"Robert, we're going to get one chance to make a point, that pen computers are a better way to manage information. If we succeed, we can change the way people work and think. If we fail, people will be lugging around keyboards from now until eternity. It's risky, but we have a shot at the brass ring."

He shook my hand and left. I felt close to him, despite our short time together. We had managed to share a very private moment. It took a lot of courage to do what he did — immediately choose to disrupt his life, quit his job, accept a position as a vice president when he had once run his own company, before selling it to Ashton-Tate. I knew right then that he was a class act, the sort of person I could rely on under pressure — a trait that would be severely tested over the coming years.

Recruiting Kevin Doren proved a somewhat more complex task. He was, after all, a cofounder of his company, lived across the country, and had recently been married. After several lengthy phone conversations, it was clear that he was interested. Persuading him was the easy part; his wife, Melanie, proved considerably more difficult.

From the outset, it was obvious that she didn't like this prospect at all. A Virginia girl in her mid-twenties, Melanie had become accustomed to Miami weather and didn't want to move so far away from her family. Kevin's strategy for dealing with her was to exhibit a high degree of patience. I followed his lead.

When she first entered my condominium on crooked Lombard Street atop Russian Hill, my fifteen-year-old cat Critter lumbered over to rub up against her. He was colored black and white in a bovine pattern, which meant that he could shed on anything and have it show. To entertain myself, I would occasionally take a black marker and alter his patterns.

"I'm not terribly fond of cats . . ." Melanie said as she sized him up. "Besides, I think he looks a bit sick."

In fact he was. His kidneys were failing, making him underweight and unsteady. I didn't expect him to last much longer.

"I hope he doesn't die while I'm here," she said.

Critter seemed to respond to this remark by staring disapprovingly into Melanie's eyes.

We spent several days driving all over town, looking at places to live — apartments, townhouses, walkups, tract houses, whatever we could find that might spark her interest. Kevin, a lanky man with a large nose and straight brown hair, was shoehorned into the back seat of my compact yellow station wagon. By the end of the second day, Kevin and I had become expert in the neighborhoods of San Francisco and the surrounding areas, having visited virtually all of them. Melanie was paying less attention to matters of geography, and after we ran out of new places to see, we discovered that we could take her back to the same ones again and get a somewhat different reaction. Little by little, Kevin softened her opposition. Maybe they didn't need a big back yard; maybe it would be easier for her to make friends if they lived in town; maybe they didn't need immediate access to a swimming pool.

On the third night, I unexpectedly had one of the most difficult experiences of my life. At about three A.M., Critter crept out of the closet where he normally slept and intentionally woke me up, which he had never done before. I opened my eyes and saw his thin, whiskered face right in front of mine. Then he went over to a particular sheltered spot under my platform bed and lay down. I got on my hands and knees to see what was under there, thinking it might be a dead mouse or some other feline treasure. He put his paw on one of my fingers, closed his eyes, and began a labored, steady pattern of deep breathing. At first he would slowly open his eyes if I called his name, but soon he appeared to be unaware of my presence. After several minutes he would abruptly stop breathing, then convulse as though slapped on the back, and begin again, like the engine of an old car that stalled, backfired, then turned over a few more times after the ignition had been turned off. Eventually he stopped altogether, and the skin on his face pulled back into a kind of grimace.

No one close to me had ever died before, much less in such a dramatic fashion. On television, people would be snagged by a bullet, then clutch their chest and collapse in a heap, but there was always a clear demarcation between being alive and being dead. This was different. The cat simply wound down, like a mechanical clock whose key was lost. To my immense surprise, I was devastated. I felt helpless and utterly alone. For a long time, home for me had been defined by where the cat was. From now on, I thought, I might as well be living in a hotel.

I felt all the worse for the cat's last act of kindness. The reason that cats seem more at peace with themselves than human beings do is that they know when they are going to die. What drives people crazy is not the fact that they will die, but that they don't know when it will happen. Furthermore, a cat's instinct when its time is up is to run away, find a quiet spot, and die alone, probably to avoid spoiling the communal den for their brethren. For some reason — perhaps the mountain of Tender Vittles I fed him over the years, perhaps those chilly days when I would let him stay warm in my lap as I worked — he decided to share his final moments with me.

When Melanie emerged from the guest room in the early morning, she immediately sensed a problem.

"Where's the cat?"

"He's gone." I wasn't interested in discussing it with her.

"For good?"

"Yep."

She walked over and awkwardly put her arm around me. "I'm so sorry," she said.

"Thanks. Sorry it had to happen while you were here."

"I know what it's like to lose someone close to you. But just remember, eventually you will feel better."

This opportunity to set aside her own concerns and connect with me somehow won her over. She soon acceded to Kevin's wishes and supported his taking the new job.

▣

Eager to get started, Kevin stayed with me while Melanie returned to Miami to arrange for the move. We got together for our first working session on a bright fall morning in my condo. Because the hill was so steep, the condo layout was strictly Alice in Wonderland: parking on the top level right next to the living room, a spiral staircase down to three bedrooms, a formal entry yet one more floor below. Robert arrived with characteristic punctuality, banging loudly on the garage door, not realizing we were sitting only a few steps away.

"Nice weather," I said as I opened the door.

"Cut the small talk, let's get down to work," Robert said in mock seriousness. He was poking fun at my own urgency to begin, even before Kevin had had time to move. As a peace offering, Robert extended a hand holding a waxed paper bag of bagels and cream cheese.

"Great, let's eat," I said as Kevin peered into the bag.

I pulled an open package of bacon out of the fridge and laid the five remaining strips on a paper towel in the microwave. "Who wants bacon?"

Kevin, always the engineer, began considering how we could divide five pieces fairly among three people.

"I know," said Robert. "That's three for Jerry and one for each of us." Kevin and I looked at each other quizzically, until we understood that Robert was referring to the founders' equity split: 3-1-1. We dived right into the discussion.

"I've done some research on components," Kevin said as he rifled through his overstuffed briefcase, finally pulling out some dog-eared papers. Robert and I had already concluded that Kevin was not the most organized person, but when it came to wild ideas, he was willing to consider anything. He smoothed out some hand drawings on the table depicting a flat, nine-by-eleven-inch tablet, perhaps an inch thick, with no distinguishing features other than a screen and a single on-off button.

"We have a balancing act to do," he continued. "Weight, battery life, screen size, and memory are all intimately related. The larger the screen, or the more memory, the more power is required. Which means more battery, which translates to either more weight or shorter life. You can take your pick."

"Well, let's look at some options," Robert said.

"I figure we can bring the product in at about four pounds, with one pound each for the display, battery, circuit board, and case."

Robert began stuffing everything in sight into my maroon portfolio, which was sitting on the table, in order to bring it up to four pounds. "Got a scale, Jerry?" he asked.

"In the bathroom, follow me." I stood and bounded down the narrow spiral staircase, which vibrated loudly. Robert and Kevin were close on my heels. We stopped abruptly and stared at the scale, realizing that it wasn't going to give a reading for something that light.

Kevin stepped on the scale and read his own weight. "Now give me the case." He took a second reading, dropping out pages until the scale read precisely four pounds more, then handed the portfolio to Robert.

Robert hefted it. "Feels heavy. I don't know if people will want to lug this much around."

"Look, the lightest laptop is about seven pounds, so I think this is pretty damn good," Kevin said, taking the matter personally. Then he turned the tables on Robert. "Anyway, it's the software that weighs the thing down."

"Huh? How's that possible?" Robert knew that software is the essence of intangibility. It is a pure expression of thought, like poetry and music, with no discernible physical form.

Kevin pointed his finger in the air, as though about to recite a passage from the Bible. "Software takes memory, and memory takes power." He glanced at Robert with eyebrows raised.

"Cost is the other problem," Kevin continued. "SRAM is expensive, about six hundred dollars a megabyte today. But I'm willing to bet that it will drop to around a hundred fifty bucks by the end of next year. So we can build the product for about a thousand dollars if we keep it to two megs. If it stays at six hundred bucks, or God forbid it goes up, we're hosed."

In contrast to software people like Robert, who are artists and craftsmen, hardware people are tinkerers and gamblers. Their challenge is to assemble things out of standard manufactured parts, as reliably and cheaply as possible. The problem is that these parts are constantly changing — in configuration, price, and availability. Hardware people must spend their leisure time poring over catalogs, price lists, and specification sheets, like professional handicappers with their racing forms. A short supply of parts instantly creates a black market, with skyrocketing prices. Brilliantly designed circuits become doomed products if a single component is unavailable. In this nerve-racking environment, hardware engineers tend to hang out together, in fear of missing the one critical word-of-mouth factoid that might save them their jobs.

Software people are a different breed. Creating software is a form of expository writing — that's why some of the best programmers have studied English or philosophy rather than math or science. Each program is like a specially crafted series of nested Chinese dolls, built out of ideas. Programmers start by carefully structuring their thoughts into tightly reasoned abstract arguments, then packaging these into simple, reusable concepts, to which they assign descriptive names. They then tower these concepts higher and higher, defining each new one in terms of the last, until they have described a complete solution — from the shuffling of bits up to familiar human operations.

"I think we should shoot for the software being one meg, which will leave one meg for the user's data," Robert offered, putting his first design stake in the ground. He turned the discussion to software.

"You know, this is a chance for us to really rethink how the user interacts with the computer. The pen interface has to be simple — you just touch the pen where you want, and do what you want to do. I think we should design the software around a physical metaphor that people can easily understand. Like a notebook, where you turn pages to go from one document to another." The volume and pitch of Robert's voice began to rise, which happened whenever he got really excited. "The cool thing about the notebook metaphor is that the user doesn't need to think about programs and files. They just turn pages to move from one task to another."

The phone rang, and the three of us looked at one another in embarrassment. We had spent the last half hour in an animated discussion, enthusiastically designing the future of computing, while crowded into the bathroom.

But the tone had been set, and now we knew we could work together. This was the last critical aspect of the deal, and everything fit together. With the idea, the money, and the executive staff lined up, all the basic resources required to start the company were now in place. We formally incorporated GO on August 14, 1987, a date we celebrated each year by drinking cheap champagne out of tiny Dixie cups.

3

The Company

COMPANIES ARE NOT conscious in the human sense, but each one nonetheless has its own personality. A company develops attitudes, even emotions, that are expressed through the actions of its people. It creates goodwill, establishes trust, and forms relationships with customers, employees, government agencies, and other institutions. This personality is captured in the corporate culture — the values and beliefs that the people in the company are expected to live and work by. The culture may be explicitly acknowledged as a credo or mission statement, but more often it exists as an unexpressed set of accepted procedures, practices, and behavior that may at times be at variance with the professed goals. This underlying code is not visible in inventories or income statements, but it is the soul of the company.

A corporation can perform acts of good and evil. It may improve the skills and knowledge of its workers, contribute to local communities, enhance the lives of its customers, advance human knowledge or international understanding. It may also stage sneak attacks on the customers or suppliers of other companies, steal trade secrets or key employees, pollute natural resources, betray the trust of partners, exploit its workers, or form alliances to gang up on a common foe. Corporate law seeks to guide corporate behavior just as criminal law seeks to guide individual behavior, with a comparable degree of success.

These characteristics are not created by fiat. They are not mandated, manufactured, or bought. They arise out of a natural process that reflects the personalities and desires of the founders, economic conditions, the available labor pool, and cultural attitudes. And that's why the first few hires are so critical.

While I looked for office space to rent, we worked out of a spare bedroom in the lower level of my condominium, overlooking the cobblestones and gardens. This was our war room, where we began the task of assembling a staff. It proved easier than we had expected. We soon developed a simple technique for selecting candidates: check their credentials, tell them about the project, and then observe their immediate reaction. We judged their suitability by gauging their level of enthusiasm.

The first person we hired was Todd Agulnick, a talented software engineer who by a remarkable coincidence had also been the first employee Mitchell Kapor had hired at Lotus — when he was only thirteen years old. Todd was now an undergraduate at Yale, but had spent the summer of 1987 working part time for me, investigating handwriting-recognition algorithms while in summer school at Stanford. He found the easygoing life in Palo Alto to be quite a contrast to dreary inner-city New Haven. It also helped that he shared a dormitory with a high school girls' cheerleading camp. There was no way he was going back east, and so he took a leave of absence from Yale to work at GO. A sensitive and fastidious young man, he was easily derailed by minor annoyances, like a disagreement with a coworker or the rumor of an organizational change.

Robert Carr turned up a promising candidate named Mike Ouye, an amiable senior software engineer who had worked at VisiCorp and later at a number of other startups. He grew up in the small farming town of Lodi, California, the son of Japanese-American parents. Mike had a moon-shaped face with Oriental features, but otherwise was as American as MTV. His Japanese heritage was to him what my Jewish heritage was to me — a remote cultural artifact whose main relevance was to add a bit of color to our personas. Mike was the quintessential computer nerd. He hung out at Fry's — the enormous Palo Alto computer superstore where techies could buy parts, software, and Jolt Cola at any hour of the day or night — and spent hours checking the ads in those thick computer tabloids available for free on street corners throughout the area.

On the day Mike was scheduled to come by the condo for an

interview, he had phoned and asked if he could bring his fiancée with him, since they were going to dinner afterward in the city.

Robert had told him it wouldn't be a problem. A couple of hours later, Mike, Robert, and I were on my living room couch reviewing Mike's résumé while his fiancée, Celeste Baranski, sat quietly reading a magazine on the loveseat. Completing the review, we turned our attention to technical matters.

"Mike, what CPU do you think we should use for the product?" Robert asked.

"Well, that depends on whether you want to stick with the Intel line or go with something else, like the Motorola 68000. As far as software development goes, the tools for the 68000 are probably more mature."

"On the Intel line, though, the problem is power consumption on the chip, which can drain the battery very fast," Robert said. "The 286 is state of the art, and it's available in a static CMOS version, which draws very little power. But the 386 will probably be available soon." At that time, the power of today's 486 and Pentium chips existed only in an engineer's dreams.

"That's true," Mike said. "But there aren't any good 32-bit compilers out yet for the 386, so you can't really take full advantage of its power. And anyway, it's probably not going to be available in a CMOS version for a long time."

"That's bullshit!" Celeste said. Robert and I swung around to look at her. "The die size of the 386 is so large that it's *only* available in CMOS, in order to reduce heat dissipation problems."

Mike felt compelled to explain this outburst. "Celeste is a hardware engineer," he said sheepishly.

She chuckled. "Damn right. And you software guys better watch it when you talk about CPUs."

It turned out that not only was she a hardware engineer, she was an excellent one, and her skills were in great demand in Silicon Valley. Apparently, she was here to check us out before deciding whether to let us know that she, too, might be interested in working with us. Within a few days, Kevin had interviewed her extensively, and quickly offered her a job.

Mike and Celeste formed the core of the technical team — he on the software side, working with Robert, and she on the hardware side, working with Kevin. They were soon married. New single employees were surprised to learn that they were husband and wife, because they continually bickered and hurled professional insults at each other during meetings. Married employees found it easier to believe.

Celeste, in turn, knew a laconic, earnest engineer named Phil Ydens, with whom she had worked at Grid Corporation, a manufacturer of portable computers. He was a specialist in the arcane area of operating systems software, the fundamental programs that move data in and out of computers and manage the allocation of resources to applications. Phil was the type of person who in previous ages would have volunteered for crusades. Tall, rugged, and steady, he lived an ascetic existence — drawn to the purity and hardship of struggling against impossible odds. He was immediately enchanted by the project, and soon signed on.

The next priority was to hire a director of documentation, who would be responsible for the instruction manuals for customers and software engineers. Robert had worked previously with an experienced writer named John Zussman, who was just finishing up a major project for Oracle, a relational-database company. He arrived for the interview impeccably dressed in a gray and blue tweed jacket, a tastefully matching tie, and polished Italian shoes. His hair was neatly cropped, his posture beyond reproach. John spoke softly and precisely, giving him a prissy air. And there we were, in our blue jeans and work shirts, looking like armchair lumberjacks by comparison.

Over the next few years, John would bring a touch of class to the organization. He arranged company excursions to the symphony, where he and his wife sang in the chorus; I organized field trips to professional wrestling matches and the racetrack. At late-night work sessions, John would dine on takeout sushi, meticulously packaged with ginger, wasabi, and a sprig of greens; the rest of us would wolf down meatball submarines.

This incongruous assortment of engineers, reminiscent of the diverse characters of a Passion play, had one thing that bound them together: a deep, almost spiritual conviction that a good idea and some hard work can make a difference. They believed in the power of a concept to change people's lives for the better, and felt a moral imperative to make it happen. GO was a vehicle for each of them to put this feeling into action, to prove to themselves that their faith was justified. Perhaps, in the end, this will be the startup game's most valuable contribution: to rekindle a sense of purpose, a sense of empowerment in a Big Brother world.

We rented a clean but funky industrial loft that had been converted to office space in the "south of Market" area. Here, commuting artists and engineers like us worked by day amid an invisible population of homeless people who slept in abandoned vehicles at night. These two communities formed an unspoken mutual interdependence: we provided them with cardboard and aluminum cans that could be quickly converted into cash, and they watched our cars and offices, which reduced vandalism. Despite the wretched conditions, the local residents eked out a semblance of decency — sharing blankets and celebrating birthdays together, with occasional contributions from us day folk.

□

Soon after we moved into our new offices, I got an unexpected call from Steve Sakoman at Apple.

"How's it going, Jerry?"

"Fine, and you?"

We exchanged guarded pleasantries until Steve was comfortable enough to get to the point. "I understand that you went ahead with the project anyway."

"Uh-huh."

"Jerry, John would like to see you."

"John who?"

"Sculley."

"What about?"

"He just wants to talk."

The intrigue deepened.

"Does he know what we're working on?" I asked. This, of course, was a loaded question.

"A little, I think," Steve said. "I've told him very little, but he probably knows the gist of it. After all, it's been a while, and I don't know what direction you've gone in. Is it pretty much the same?"

Now Steve was fishing. I dodged the question. "Hard to say."

"Well, anyway, he'd like to see you."

I consented to the meeting, and we set up a time.

On the day of the meeting, I climbed over a pile of unidentifiable soggy trash blocking my car and drove the hour down to Apple headquarters in Cupertino. The company was housed in a series of sparkling beige buildings adorned with colorful Apple logos. I parked near a particularly airy edifice of stucco and glass that had a spectacular view of an artificial pond through the main lobby. Registering with the guard at the desk, I felt as if I were checking into a spa in the Emerald City of Oz.

Steve appeared moments later to greet me. He was all sunshine and smiles.

"Hi, Jerry, how have you been?"

"Fine, and you?"

He led me through endless corridors of pastel-colored workstations, each equipped with the latest in user-friendly Macintosh technology. Smiling icons and desktop-published party invitations were pinned up everywhere.

"This is my group's area," he said, waving his arm vaguely over a large expanse. He walked around at a pace designed to allow me a glimpse of their project without learning anything important. Hung on a wall was a long poster from an elementary school instructing children on how to print the alphabet correctly. I glanced at one workstation long enough to see electronic ink on the screen, but little else.

He saw me pause for a closer look, and he took my arm and said, "Maybe we should get going. I don't want to keep Sculley waiting."

We returned to the front entrance and started toward Sculley's office. We were barely out the door when we were nearly tackled from the rear by a rough-and-tumble, athletic-looking middle-aged man with a raspy voice.

"Sack-O-Man!" he said with the energy of a college quarterback.

Despite his years, I could tell he was someone who had always looked younger than he was. He grabbed Steve and gave him a big hug. Steve didn't seem to object. I couldn't imagine what would cause a grown man to exhibit this kind of exuberance. Without warning, he turned his attention to me.

"Jerry Kaplan! How are you, boy?" He looked familiar, but I couldn't place him. I knew I must have met him somewhere before, since he knew my name.

Steve bailed me out. "Jerry, this is Bill Campbell, president of Claris." Claris, a spinoff company created by Apple, was developing applications for the Macintosh. Campbell shook my hand vigorously while Steve finished the introduction.

"Nice to see you technical geniuses looking so fuckin' dapper!" Campbell laughed as he vaulted off to his car.

"We're going to see Sculley!" Steve called after him, by way of explanation.

"That's your problem!" Campbell shot back.

As he screeched out of the parking lot, Steve said, "He's got his own style for sure, but he's really a great guy. He'll run this place eventually." This was high praise coming from Steve, particularly for someone from a marketing background.

We arrived ten minutes early at Sculley's office. The first thing I noticed was that there was not one, but two administrative assistants there to handle his affairs. The furniture was classic California hip, unpretentious but no doubt very pricey — rich light woods, polished to a warm glow, projecting a calculated blend of limitless resources and political correctness. The reception area was enormous compared with Sculley's office — visible through a glass partition — which seemed to be built for the purpose of showing off the furniture.

"Please wait here, John is just finishing up with his personal publicist," said the more senior assistant.

I whispered to Steve, "Personal publicist? Why the hell does he need that?"

Steve smiled bashfully and shrugged.

Jean-Louis Gassée, the fellow at Apple who had talked Steve into staying, instantaneously appeared as if out of thin air. He must have been watching me eye the place. He leaned toward me and said, "Obscene gross margins."

"What?"

"This splendor is the happy result of Apple's obscene gross margins. And do you know why?" He spoke with a smooth French accent. "Because we control the whole platform, both the hardware and the software. The rest of the PC manufacturers are going to the dogs, eating each other for lunch as they get squeezed by Microsoft on the software side and Intel on the hardware side." He gestured out the window, to the world outside Apple. "Would you believe they want us to license our software to others?" He narrowed his eyes at me, as though I should know precisely what fools he might be referring to. "But we don't have any competition. I say keep the prices high and the profits will roll in."

"What about Windows?" I asked. Windows was the graphical user interface developed by Microsoft in an attempt to cut into Apple's market. The strategy was to copy the Macintosh's "look and feel" onto IBM PCs and compatibles. To date, this approach hadn't worked. Windows 2.0 — the second attempt by Microsoft — had just come out, and was getting more reasonable reviews, though it was still a far cry from the polish and ease of use of the Macintosh. Bill Gates had started the project almost immediately after Steve Jobs had privately shown him the first Macintosh, as yet unannounced, in order to persuade him to write applications software for the machine. Gates had been impressed — he had an unerring eye for important new technology — and agreed to write applications for Jobs, but with a deeper goal: to learn enough about this new graphical technology to take the market away from Apple. Now Gates's company was a leading supplier of Macintosh applications, as well as Apple's leading challenger. Gates had them right where he wanted them — dependent on him as a competitor.

"Windows is a joke," Gassée said. "They'll never catch up." He looked approvingly at Steve Sakoman. Even for an Apple executive, Gassée's complacency seemed extreme. Nonetheless, it was fed by an ever-widening circle of Macintosh devotees and sycophants. Like Pepsi, where Sculley had been an executive, Apple was selling a product wrapped in a dream: buy a Macintosh and it will make you cool. As in sports car ads, the models in Apple's advertising tacitly posed the question: Do you want to look like me? Buy this product and you will.

Sculley was the perfect CEO for Apple: greatly concerned with corporate image, not to mention his own. Under the paternal eye of its founder, Steve Jobs, Apple had reared an entire generation of yuppies who had ripened from innocent yearning into rampant materialism. This was Jobs's legacy — imprinting his self-interested personal style on thousands of impressionable young Apple recruits, who tended to measure their success by job title and the number of people below them on the organization chart. In a symbolic rite of passage, Sculley had deposed the creative and irascible Jobs in a palace coup, replacing Jobs's persona but not his technical insight. The organization adapted accordingly.

Sculley opened his door and invited us to enter. He had a chiseled, windswept look, and his head sloped back as if he were gazing far into the distance. An improbable wave of brown hair completed his young statesman's visage. Gassée ushered us into the room as though he were the host, and followed us in. I surmised that he acted as a kind of gatekeeper to Sculley, at least for certain matters.

Sculley's office was dominated by a round wooden table supported by thin chrome bars clamped to a chrome rail. It gave the illusion of hovering in midair, defying gravity. He sat directly opposite me. Steve was to my left and Gassée to my right, with hands cupped on the table as though about to play a round of poker.

Sculley began. "I'd like to show you something. This hasn't been seen outside Apple yet, so I have to ask you to keep it quiet."

"No problem," I said.

He swung around and opened the doors of a cabinet. Inside was a complete audio-visual setup. He inserted a videotape and the screen lit up.

An actor portraying a pleasant but nerdy-looking fellow in a white shirt and suspenders was sitting at a desk. The room was filled with knickknacks of various styles and eras. A rotating ceiling fan projected dramatic shadows across the scene. On the desk in front of the actor lay a flat, art deco tablet with a screen. As he spoke, the screen came to life.

"I need to give a lecture on the African rain forest. What's available on this subject?" The actor was obviously supposed to be a professor. A small "window" opened on the tablet and a smiling geeky face with a bow tie appeared, eager to please.

"Just a moment," the electronic pixie said. "I'll check all major libraries." In seconds, a bibliography appeared on the screen.

For the next several minutes, the actor engaged this electronic elf in a free-ranging dialogue. The elf displayed a mastery of colloquial English and conversational dynamics. Periodically, the actor would touch the screen to bring up startling color pictures, answer a video phone call, or select from a menu. The drama ended on a light note, with the actor sneaking out when a call from his mother appeared on the screen.

At first I was a bit confused. What was the point of this fantasy? My background was in computational linguistics, the processing of human languages by computers. Knowing the practical difficulties of speech recognition made me less able to get into the spirit of the thing. To me, the premise of this slick, expensive production was preposterous.

When the video ended, Sculley swung around to assess my state of mind. "What do you think of our Knowledge Navigator?" This was apparently the name of the tablet sitting on the desk, which existed only through the magic of video special effects. To his dismay, my feet were firmly planted on the ground.

"Are you planning on building this thing?" I asked.

Before he could answer, Gassée, noticing my skepticism, jumped in. "Well, it's really just a concept piece."

"Yes, it's our ten-year vision, the direction of our company," Sculley added.

I wasn't going to let him off that easy. "John, since the ancient linguist Panini, who in the fifth century B.C. attempted to write the grammatical rules of his native Sanskrit, scientists have been stymied by the complexities of language. I don't think Apple's engineers will solve this problem in the next decade." Steve Sakoman winced noticeably. In retrospect, I had probably sounded too pompous.

"As I say, it's really just a concept piece." Gassée was clearly doing damage control. He seemed as concerned that I might raise doubts about their project in Sculley's mind as he was with my own reaction. But Sculley wasn't giving up.

"Maybe we can't get there, but we can sure give it a try. As you know, Steve is working on a secret R&D project to bring the Knowledge Navigator to life, at least in an early form." I wondered what exaggerated visions of Steve's project were dancing in Sculley's head. "In the early days of the Macintosh, we invited in selected software publishers to work closely with the development team. We hold you and your team in high regard, and would like to offer you the opportunity to be the premier applications developer for Steve's project."

Gassée instinctively placed his hands flat on the table, as though he had just laid out his cards. Now all was clear. Sculley was using a time-honored technique with me that he had seen Jobs use repeatedly to his benefit: get a techie in the room, wow him with a slick personal pitch and a prototype, then hope he invests his time, money, and talent in support of an Apple project.

I was so amazed at this ill-conceived attempt to sidetrack GO's project that I couldn't think of an appropriate response. I decided on the cowardly tactic of laying it off on our backers. "I appreciate the offer, but I've just raised money on a business plan that has specific goals, milestones, and deliverables, and the VCs aren't going to stand for me changing horses in midstream." This was only partly true. There wasn't much of a business plan, and the backers would probably have gone along with Sculley's plan if we had asked, but I wasn't about to flake out on my teammates, especially in response to a magic show.

"Perhaps you could consult with them," Sculley politely suggested.

"John, I'll be happy to explore the issue with them and get back to you if they show any interest."

I turned to Steve as we walked out. "Thanks, Steve, it was a nice thought." I think he was a little embarrassed.

"Yeah. And good luck to you, Jerry."

I knew that he meant it. We were athletes in the same sport, just playing for different teams. We both knew that we had a lot in common, and that it might well be different the next time around. There was no reason to burn our bridges.

A few years later, Steve Sakoman left Apple to join Jean-Louis Gassée in a new venture, Be Labs. I don't think I ever talked to him again, but after his departure, his project went on to become the Great White Hope at Apple — the Newton "personal digital assistant." It was destined to have a profound impact on GO.

When news of the Apple meeting spread through our offices, it created a good deal of excitement. Perhaps it was the thrill of knowing that someone of Sculley's stature was worried about us, or perhaps it was the knowledge that there were real competitors out there. Whatever the reason, this first external threat to our company galvanized the team. It was time to establish a more explicit corporate culture and set near-term objectives. I discussed the subject at length with Robert Carr and Kevin Doren, then called a company meeting one day after lunch. We gathered in a small open area of our office near the Pullman-style kitchen.

I had passed the word around that this might be a long meeting, so the first order of business was for everyone to stock up on munchies before we got started. There was always plenty of free food in the kitchen. The rule was that any food not labeled could be eaten by anyone — "No name, fair game." The refrigerator shelves were dotted with little yellow stickies.

Because I often worked late and was willing to eat anything, it soon became a matter of sport to feed me. I had a strict nondiscrimination policy with respect to color, age, or ethnic origin in my diet. That day, Celeste had put a special sandwich in the fridge with my name on it. This was an unusual courtesy, one that seemed too good to be true. Biting into it, I discovered two tiny pink plastic pigs, which I pocketed, and consumed the rest of the offering. Of course, she had told everyone about the prank, and they all waited for my reaction.

"Hey Celeste, great sandwich. And thanks for the pig, too!" I called out across the room.

"Pig?"

"Yeah, the little pink piglet."

"*One* piglet?"

"Why, does he have a friend?"

She sat down and turned purple, to everyone's delight.

I called the meeting to order. "It may be hard to believe today, but pretty soon everyone in this room is going to be an old-timer at GO. We're the ones that new recruits will look to for guidance, to see how they should think and act. Today I want to talk about some principles I'd like us to work by.

"When I was a kid, I thought the moon shot was a remarkable scientific achievement. I now understand that the true triumph was not only technological, but also managerial. It was in 1961 that President Kennedy set a goal of putting a man on the moon by the end of the decade. It was in 1969 — only seven years later — that this goal was achieved."

"Eight!" At least four people yelled this at once.

"Nine minus one is eight," Kevin said softly.

"Thanks, I'll remember that. During that time, hundreds of thousands of people were mobilized into a carefully choreographed ballet. There was no room for egos, no place for bullshit, and no excuse for delays. People's lives and the prestige of the country were at stake.

"Now, our task is simpler, but it's still very risky. We are building an unproven product for an unproven market. And the key to success is to reduce risk whenever and wherever we can. Every one of us has an important role to play, and that makes us all part of management. In our company, everyone is responsible for understanding our goals and how their work contributes to it. That's the first principle. If you don't understand what's going on, you should always feel free to raise your hand and ask."

Every hand went up in unison, and the room exploded with laughter. I shook my head in mock disgust.

"The second principle is that there is no such thing as a great product that doesn't sell. We aren't here to promote truth and beauty, but to meet customers' needs.

"Last, we treat everyone outside our organization with courtesy and respect. Here's why. We have to innovate in hardware, software, applications, communications, marketing, and God knows what else. This project is so big, we can't possibly get enough money to do it alone. We need other people to invest their attention and resources right alongside us. So here's the strategy: we should do just enough in each of these areas to interest competent partners. If they're willing to step in, we should let them, and we'll back off. We have to provide compelling economic opportunities for everyone."

"Now wait a minute," Todd Agulnick said. "Suppose we give everything away like you suggest. Then what's left for us?"

"Yeah, are we gonna wind up flipping burgers at the Doggy Diner?" Celeste added, gesturing toward the all-night burger joint on the corner, topped by a giant plastic smiling dog wearing a chef's hat.

Robert broke in with the answer. "Todd, when you strip everything else away, our value is the APIs."

API stands for application programming interface, the building blocks of the internal operating system out of which application programs are constructed. Every operating system has its own APIs that define what the system can do, how it will look to the user, and how to make it happen. APIs are the foundation on which applications programmers build and on which their products must depend.

Creating an API is like trying to start a city on a tract of land that you own. First you try to persuade applications programmers to come and build their businesses on it. This attracts users, who want to live there because of all the wonderful services and shops the programmers have built. This in turn causes more programmers to want to rent space for their businesses, to be near the customers. When this process gathers momentum, it's impossible to stop.

Once your city is established, owning the API is like being the king of the city. The king gets to make the rules: collecting tolls for entering the city, setting the taxes that the programmers and users have to pay, and taking first dibs on any prime locations (by keeping some APIs confidential for personal use).

This is the great secret of the computer industry. While the newspapers chronicle the daily skirmishes of computer companies and their products, the real wars are over control of APIs, a subject too arcane to attract the attention of the press. But once an API is established, the owner has a natural monopoly of unprecedented proportions. API wars are brutal, winner-take-all conflicts in which the losers become mere shadows, eking out a marginal existence in specialty markets.

I completed Robert's point. "It only looks like we're a product company. In the end, the real game is won by establishing our own APIs. Because pen computing is a new industry where no one is established yet, we have a chance. If we succeed, it will be damn near impossible to stop us." I waited a moment so the significance of this seeped in.

"I want to do one last thing before we break. In honor of these principles, I would like to propose a company mascot — the ant."

I pulled an ant farm out of a paper bag. Not just any ant farm, but an Uncle Milton's Ant Farm. This delightfully nerdy toy, touted for generations in the back pages of comic books, came with its own ants — real live ants, delivered by mail in a test tube. The newly installed residents had already set about constructing tunnels, which could be viewed through the plastic panels.

"Yuck-o!" exclaimed Todd.

"Anyway, I think they capture the right attitude. They are egoless, cooperative, and diligent, and are excellent engineers. Also, they live on a few drops of sugar water a week."

Todd was mentally exploring the possibilities. "Can we watch 'em breed?" he asked.

"We need to get you a girlfriend," Celeste said. He grinned and gave her a thumbs-up.

As the meeting broke up, Phil Ydens — the laconic operating systems engineer from Grid, who had sat quietly throughout — walked up to me and said only one word, "Cool." Then he returned to his desk. I took this as an expression of his approval. It made my day.

Kevin, Robert, and I were the only ones left in the common area. The mood had been light, but the three of us knew the seriousness of the situation. We had spent several months now investigating the project, and it wasn't going to be a cakewalk. There were all kinds of risks and no guarantee that we could make the milestone we had promised the investors: to demonstrate a deskbound prototype of a pen computer by June of 1988, only about six months hence. Kevin was the first to speak. "GO is destined to be spectacular for sure — one way or the other."

"Yep, we're either going to be a spectacular success or a spectacular failure," Robert added.

I bent over to pick up my empty soda can. "Either way, it's going to be a hell of a joyride."

Kevin raised his eyebrows and nodded. "Or a joyride to hell."

For the next several weeks, the ant farm was the centerpiece of the office. Feeding time was a major event, and since everyone wanted to do it, we had to put up a schedule. One day I arrived at the office to find a sullen group standing around the farm.

"Why's everyone so down? This place looks like a funeral."

Robert pointed to the ant farm. "It is." Our cherished mascots were now just curled black specs, lying motionless in their carefully excavated home.

Todd was especially upset — he had developed a strong personal identification with the ants. He said, "I hope this isn't going to happen to us."

After cleaning out the ant farm, I wiped the last of the sand from my hands and walked back to my office. I encountered Kevin in the hall. "Joyride to hell," he repeated.

As we got deeper into the project, it began to consume more and more of my time. Soon there was little left for any personal life. Before starting the company, I had been dating a flight attendant from Sweden, but she had grown increasingly impatient with my preoccupation with work. My parents also sensed a problem — it would sometimes take me a week to return their calls. They had only the vaguest understanding of what I was doing, and grew suspicious that I was risking my nest egg on a foolhardy undertaking. Unable to satisfy their curiosity by phone, they finally decided to make good on their threat to visit.

Mickey and Murray Kaplan were an upper-class New York Jewish couple whose Depression-era mindset had slowly evolved into an unlikely blend of political liberalism and a deference toward those with more money than they had. Art lovers without truly deep pockets, they had parlayed my mother's unerring eye for promising young artists and my father's obsession with bargains into a respectable collection of modern art. Drawn to the competitive and all-consuming Manhattan social scene, they trundled their four children off to different private schools. My siblings and I were free to take lessons in anything we showed an aptitude for, and encouraged to compete with one another for grades and honors. Predictably, we went our own ways, making separate friends and adopting different values.

Like many of their acquaintances, my parents spent winter holidays in Palm Beach, Florida, an insular enclave of face-lifts and hearing aids where the most frequent social gathering was the funeral. New York Jews must be descended in part from migratory birds, for as their children leave the nest, year by year they lengthen their winter visits south until they never return north. My parents eventually crated up their art collection and swapped their luxury co-op on the East Side for a spacious classic-revival home near the quiet north end of Palm Beach Island.

Arriving at the airline gate to pick them up, I noticed that my mother looked younger than I had remembered her. I couldn't be sure if this was a natural condition or the result of some airborne cosmetic miracle. My mother, a walking testament to the wonders of modern medicine, looked about the same at age sixty-four as she had at forty. My father, in contrast, looked all of his eighty years, but inexplicably had the rugged good looks of a seaman, though he hated boats.

We waited for what seemed like hours at the baggage carousel. Although they were coming for only three days, they brought two enormous suitcases plus four carry-ons.

"Jesus, what's in this thing?" I said, straining to lift my mother's case over the lip of the carousel.

"She brought her workout weights," my father quipped.

As we pulled out of the parking lot, they got down to the purpose of their visit — giving me the dreaded parental third degree: Is it always so cold here? Are all the highways so clean? Do you always dress this way? How come you're wearing a sweater and not a coat? How much stock do you own? Are you selling too much to the investors? When can we see the product? How much will it cost? Why does it take so long to build? What will happen if you can't pay the financing back? Where do you park? Why don't you buy a new car? Do you need any money? Are you hungry? Are you cold? When we arrived at the condo, they went to wash up, so I got a brief break from the interrogation.

Before unpacking, my mother headed for the refrigerator. Though my parents were reasonably trim, they seemed to eat as often as newborns. At first I thought she was hungry. Then I thought she was checking to see what I kept there. It turned out that she was making a deposit. Whenever they traveled, she brought along a variety of items in small plastic bags: carrot and celery sticks, lurid orange cracker sandwiches filled with processed cheese, some crumbled cookies, and slices of apple that had long since oxidized to a purplish brown.

"Why do you bring this stuff?" I asked.

"You never know when you need food. Your father gets hungry.

I traveled all through Europe with a bag of food. In Russia, I took enough on the train to feed the Russian army. Fortunately, the Russian army was on the train, but when I tried to feed them, they wouldn't take any. I think they thought I was trying to poison them."

Wise move, I said to myself.

The next morning we visited my office. After a tour, a show-and-tell, and some introductions, my mother wanted to make a copy of a recent article in a trade paper that contained a picture of me. "Mother, someone is using the copier. How about if I do it later?"

"Can't you just kick 'em off?"

"No, they're doing real work."

"But you're the boss. They have to do what you say."

"That's not the way things are anymore."

The office manager, who had overheard the conversation, came to my rescue. "I'll take care of it."

"Thanks." I handed it off. My mother went with her, happy to have a fresh face to put questions to. This gave me a few minutes alone with my father.

"I can't believe you quit a real job to go do this. They'll skin you alive if you can't pay the money back." He spoke in an urgent whisper, leaning in as though to warn me.

"Look, it's really OK, Dad. I'm not going to starve. This isn't the rag trade." "Rag trade" was the slang term for the New York garment industry, where my father got his only real business experience.

"You mean they didn't make you sign for everything?" He was referring to the practice of personally guaranteeing a business loan, as he had been required to do each season to finance his inventory, before his business went sour.

I laughed unintentionally as I realized just how much things had changed since his time. Back then, starting a business meant risking everything — your savings, your house, your self-respect. You *were* the business, and if it failed, all was lost. But today, the distinction between the business and the individual is much stronger. The individual is legally insulated from personal disaster, a circumstance that encourages risky projects like GO. I wondered briefly if I would have had the nerve to start the company forty years ago.

He turned his attention to office security. "I notice that people came in before you did this morning," he said.

"Yep, the place is open pretty much twenty-four hours. We have a kitchen and a bunch of futons in case people want to sleep."

"Do your partners watch the workers when you aren't here? Can you trust them?"

"Look," I explained, "everyone is a partner. Everyone's got stock. They all keep their own hours. If they steal something, they're stealing from themselves." He looked totally confused. "Think of it like partners in a law firm."

This was better. Now he merely looked skeptical. He changed the topic to my personal life. "It looks like you're having some problems with your girlfriend."

He was right. "It's hard on her. I have to work all the time, sometimes seven days a week. She loves to go out, to get away, but we can't plan anything." He put his hand on my shoulder reassuringly as I spoke. "The project's on a tight schedule, and I just can't take any vacation until we get the prototype running. When I come home I'm exhausted, and she wants to party." I was starting to feel sorry for myself.

"I know, it's a real problem," he said sympathetically. "When you're in the middle of something, it always seems very important. It's hard to take time for personal matters, to take a deep breath and enjoy the human side of life, to build meaningful relationships with your friends, to learn to really listen."

"Yeah, I know. And it's driving her away."

He gazed out the window as he collected his thoughts. Then he spoke. "You've just got to explain to her that work comes first."

I couldn't believe he was serious, but he was.

"Look, women come and go," he explained, "but in business *you're playing for keeps*."

When we left for the airport three days later, my mother collected her plastic bags from the fridge and put them back into her purse. We got to the airport a full forty-five minutes before flight time — long before any self-respecting business traveler would consider arriving. As the gate attendant called their row, my father turned to me and said, "You know, your mother and I are very proud of you, no matter what happens. We know you'll be all right eventually. Always remember one thing. You got a good education, and that's the key. They can take away your money, they can take away your business, they can take away your house, but they can't take away your brain."

Over the years, he often said this when we parted — right before handing me a wad of bills from his wallet.

▣

By early 1988, the office had taken on the feeling of a city hospital after an air raid. It seemed as though we'd uncover a new and more dreadful problem every day.

It turned out that flat-panel displays couldn't withstand the pressure of writing on them with a pen. Protecting them with a separate sheet of glass made the electronic ink appear disconnected from the tip of the pen, particularly if you moved your head to different viewing angles. The glassy surface was so smooth that writing on it felt like roller-skating on ice. If we etched the surface to add some resistance, it muddied an already hazy image.

We confronted equally daunting obstacles with the software. There had to be no perceptible delay between the time when the pen touched the screen and the electronic ink appeared. This meant circumventing the normally sluggish input-output processes and writing our own. The available pen sensors would go out of alignment at the edges of the sensing area, so we wrote correction routines. However, these had to be specially calibrated to each individual prototype, and even on a single unit the calibration would drift as the electronics warmed up. And worst of all, the handwriting recognition was awful.

It seemed as if we spent all of our time solving problems that had little to do with the goals of the project. Each day, the engineers would go home to nightmares about millimeters of pen-sensing distance, white-to-black pixel transition times, and dropped software interrupts.

A sense of panic set in. We had promised the backers a working demo by the end of June, but were wasting months on things that we had taken for granted in our schedules. I called a meeting with Robert and Kevin. They shuffled into my office with fear in their eyes, like schoolboys summoned without explanation to the principal's office. But they knew what the problem was.

"Look, guys, let's face it. We're heading for trouble," I said. "So why don't we approach this differently? Let's redefine the problem as one that we can solve. For June, we can rig up a stationary demo so that the ink always appears under the pen tip. We can downplay the importance of handwriting by showing off how we can manipulate electronic ink and graphics."

On the whiteboard in my office, we did a quick inventory of what was shaping up well and what wasn't. We then reset the goals of the promised demo to showcase the good stuff. As if by divine providence, Mike Ouye stuck his head in the door. He looked relieved to find the three of us together.

"I've got something you should see."

He led us to his cubicle without saying a word. When we arrived, the entire staff was standing around solemnly, like ushers at a wedding waiting for the service to start. They parted to let the three of us stand by Mike's computer. On the screen was some text — the opening paragraphs of Jack Kerouac's *On the Road*. Mike handed me his electronic pen. "Try it," he said.

I carefully drew an X over the word "hitchhiking." It disappeared instantly, and the surrounding text closed up the gap. "Son of a bitch," Robert said. Then I tapped on the word "highway" and dragged it several lines down. Like magic, the paragraph reformatted itself to reflect the new order of the words.

I was speechless. For almost a year, these actions had existed only in our minds. Now there they were, and they were beautiful. It was like imagining how it would feel to fly and then getting the chance to actually do it. The feeling of immediacy and simplicity that we had speculated about was right there on the desk for anyone to experience. And it worked.

Our earlier panic turned into a growing sense of urgency. From that point on, the office was almost never empty. People would work around the clock, sleeping on futons and eating takeout from Doggy Diner. In the remaining months before the June deadline we managed to cobble together enough of a demo to convince even the most hardened skeptics of the potential of pen computing — except for one serious remaining problem: the handwriting recognition was abysmal. I felt personally responsible, because in principle this was my area of expertise. I had worked closely with Todd Agulnick and Mike Ouye on our recognition algorithms. And now there was less than a week to go before we were to demonstrate the results of our labor to the board of directors.

"I notice that everyone writes the same thing," Todd said. "'Four score and seven years ago.' Then they try their name. Maybe we should just figure out a way to distinguish these and a few other key phrases."

It was a sign of our desperation that we momentarily considered something this ridiculous, but we were so concerned that no idea was rejected out of hand. Meanwhile, Mike was trying to track down some last-minute glitches in the pen-sensing software. He wrote a program to plot the pen position immediately on a graph whenever it moved. We watched the screen.

"What's that?" Todd asked, pointing to a sudden spike in the graph.

"Dunno," Mike said. "Seems to happen sometimes."

We looked at one another and instantly realized what the problem was. "Mike, you asshole," Todd yelled. "Those are bad points coming from the pen sensor!"

"Holy shit, you're right."

The pen sensor was occasionally reporting wildly incorrect positions for the pen. In the middle of a smooth, continuous stroke, it would indicate that the pen had skipped to some remote spot on the screen, then mysteriously reappeared near its original position. This would screw up the recognition process, which was lacking any common sense about pen motion, and so would try to interpret this abrupt movement as part of the writing.

It was a simple matter to screen out such points in the software, since they were physically impossible movements and easy to identify. To everyone's delight, the recognition improved dramatically. We all took turns marveling at the software's newfound ability to recognize our words written in block letters, though it was still far from perfect. Then we put the demo to bed and went home to do the same.

4

The Financing

In business, as in life, major passages are steeped in customs and rituals. There is, however, one significant difference: religious rites express the emotions of the participants, while business deals obscure them. This is not because business is devoid of powerful feelings — on the contrary, it is driven by them — but because the dominant sentiments are greed and fear rather than love and joy.

Venture capitalists are masters of procrastination. Their standard practice is to keep all options open until the last possible moment, often abandoning some hapless entrepreneur at the altar after months of reassurances that there is just one final question to answer before they will invest. The VCs know that time is on their side: the less money a venture has, the more desperate it will be to make a deal. They stand on the sidelines with their hands in their pockets, like bystanders afraid to get involved, while small companies slowly bleed away their precious remaining cash. This is why frustrated entrepreneurs sometimes refer to VCs as vulture capitalists.

The second and third rounds of financing are often the hardest. Although everyone involved would like to think that the process is rational, in practice it is as subjective as buying a house: it's tough to pony up big bucks unless you are really fired up about something, and after all, who knows what it is really worth? But the first-round backers have stolen the excitement of finding a new, virginal deal, and the euphoria of cashing out may yet be years away. As a result, trying to corral investors for an intermediate round is like herding cats.

First the current investors have to commit themselves to participating in the new round. This shows that they remain convinced of the merits of the venture. However, they cannot credibly set the price, because they are usually interested in "writing up" the value of their portfolio to improve the rate of return they can report to their own investors. Therefore the price must be established by a new investor, called the lead. The lead is responsible for negotiating terms with the company. Then, in order to spread the risk — and give the lead investors confidence that they aren't going to look like fools if the company goes bust — the deal is syndicated to the other interested parties that the company has been courting.

In order to consummate a deal, the company must repeatedly perform the dance of a thousand veils. Potential investors must be independently enticed, coddled, and reassured by the company, which has to hold out a realistic threat that the deal may go down without them. The trick is to get each investor to "soft circle" — venture capital talk for conditional approval of the deal — and only then to introduce him to the other investors. If the introduction is made too early, the assembled group of waiting investors gets restless and colludes to bring the price of the deal down. If the introduction is too late, a lone investor may conclude that no others are interested, and therefore lose interest himself. The VCs are always trying to wheedle out of you who else is interested; you're always trying to keep the parties separated until the right moment.

By the time of the demo to the board of directors in mid-July, we had already cannibalized most of our conference rooms for offices and labs, so the meeting had to be held in a small corner room originally intended as a storage closet. The space was dominated by a large diagonal steel girder, painted burnt orange to forestall rust. This tasteful architectural detail was supposed to keep the floors from collapsing in case of an earthquake. But in that event, the turn-of-the-century red-brick façade would likely fall away, converting the structure into a full-scale living doll house. As a show of respect for the board members, our office manager put a small plastic vase holding three daisies and a paper plate of fudge brownies in the middle of the narrow, folding conference table. She stood guard against hungry poachers until the meeting got started.

I sat at one end of the table, surrounded by stacks of handouts describing our progress and financial condition. Since I was trying to diet, I pushed the plate of brownies just out of reach, to reduce temptation. John Doerr and Vinod Khosla sat on one side, Robert and Kevin on the other. Mitchell Kapor was the last to enter. He wedged his chair into place between the door and the girder. He stared longingly at the plate of brownies, then slid them away from himself, back toward my end of the table.

I passed around the first handout, a single sheet of paper. "In order to help structure our discussion today, I put together a short agenda."

Mitchell grinned. "Let's cut the bullshit and get to the demo."

"Yeah!" Robert seconded the motion.

I glanced around the room to assess the group's sentiment. They looked like fraternity brothers waiting for the treasurer to end his report so they could start the porn flick. John bolted toward the door, which of course was blocked by Mitchell's chair. Everyone stood up as Mitchell and John struggled to lift the chair out of the way, banging it loudly back and forth between the door and the girder.

Several engineers were hovering outside, curious about the proceedings and wondering how long it would be before the demo. I had told them that the meeting would probably go on for about an hour before the board moved to the demo room. The crowd grew when the door flew open, attracted by the commotion. It sounded as though a brawl had broken out the moment the meeting started.

We trooped down the hall to the hardware lab, with Robert leading the way. He sat down on a high stool in front of a workbench. On the bench, an electronic tablet was fastened to an angled lectern, with a flat-panel display carefully taped down on top of it. Wires ran behind the bench to where the computer was hidden. The board members gathered in a semicircle behind Robert; the rest of the office — by now following the crowd — flowed into the room to watch the show.

Robert was renowned for giving good demos. This stemmed from an unusual skill: he could deliver a relaxed and articulate exposition while he surreptitiously carried out a complex sequence of operations with his hands. He held your attention with his words as he deftly recovered from problems that would normally require total concentration. If it weren't for his interest in computers, he probably would have become a magician.

He described the setup, then explained how it could be fitted into a package about an inch thick. He weaved a tale about a traveling salesperson, chronicling his day and illustrating how the pen computer could improve his productivity at every turn. Although the assembled crowd was familiar with the details, they had never heard them expressed so eloquently. When Robert finished, they broke into spontaneous applause.

Ecstatic at seeing this progress, the board members returned to the meeting room and engaged in a spirited discussion about the next steps. "How quickly can you build a portable model?" John asked.

Robert, Kevin, and I exchanged glances. We were becoming adept at the type of silent communication that develops between long-married couples. "I'd guess we could have it by next June — about a year," I said. Robert and Kevin nodded.

Vinod challenged us. "What can you do to speed that up?"

I turned it around. "Help us to raise capital. At our current rate, we have about four months of funds in the bank — six, tops. Then it's 'game over.'"

"Financing won't be a problem," John said. "Everyone will want in on the deal."

"At the right price," Vinod cautioned.

Then Mitchell spoke up. "We shouldn't let a shortage of cash stand in the way of this remarkable project. Jerry, what was the price of the last round?"

In addition to the original investment of $1.5 million, we had done a smaller round of $500,000 three months later from the same investors, after we had a chance to get our budgets in order. "The original investment was at forty cents a share, the next was at sixty," I said.

"And with the shares issued to employees since then, what's the current valuation?"

"Let's see . . . Sixty cents times about ten million shares . . . About six million."

Mitchell thought for a moment. Then he said, "I think we should go for twice that. Twelve million."

Robert and I looked at each other in disbelief. Was the demo so compelling that it doubled the value of the company? We turned to John, expecting him to calm Mitchell down. John slumped forward, putting his head down into his cupped hands on the table. I assumed he was composing a polite way to tell Mitchell he was full of shit. John sat still for several seconds, then started to bounce his left leg on the floor like a fidgety schoolboy. He abruptly shot upright.

"I think we should ask for sixteen million." John had tremendous respect for Mitchell's judgment, but he was accustomed to having the most extreme view. He was not about to be out-enthused.

Mitchell and John now engaged in a staring contest, as though one card sharp had just called and raised the other's high bid. Kevin began to slide lower in his chair, as if to get out of the way.

I thought I should do something to cut the tension. "Who do you think we ought to approach?"

Mitchell answered. "Look, we need the support of the major ISVs, and they're all flush with cash. Let's approach them." These were the independent software vendors, companies that write applications that run on other companies' operating systems. He was talking about Lotus, Microsoft, and Ashton-Tate — the big three.

John frowned. "You're not considering talking to Microsoft, are you?"

Knowing Mitchell's long-time rivalry with Bill Gates, I wanted to derail this line of thinking as quickly as possible. I responded directly to John. "We have to clue them in eventually, since we want our platform to be open to all applications developers."

"He'll just start a project right away to compete with us!" John said.

"Not if we get him to invest," I said. "And besides, this way I can get him to sign an NDA" — a non-disclosure agreement, a legal document that restricts what the recipient can do with information that he learns from you. It allows you to disclose confidential information on a limited basis while still protecting your intellectual property.

John's frown slowly changed to a smile. "That might make sense if the price is right. After all, we're giving them early access to our stuff." He rubbed his hands together briskly. "How about *twenty million?*"

Kevin looked seasick. He leaned his head back against the wall, his mouth ajar.

"OK with me," Mitchell said.

Vinod, who had listened impassively to the proceedings, decided it was time to offer his opinion. "Look, there's no harm in asking, but this is a dangerous game. Pricing of these intermediate rounds is highly unstable. If they think you're close to running out of cash, they'll wait you out. If the price starts to crack, everyone gets cold feet."

John bristled at Vinod's gloomy tone. "Hey, there's no right price. It's willing buyer meets willing seller!"

"I'm not saying don't do it," Vinod said. "I'm just saying it's dangerous to start too high."

After the meeting broke up, Robert, Kevin, and I assembled in my office for a post-mortem. "I guess that went well," I said.

Kevin still looked queasy. "Too well. These sky-high valuations give me a nosebleed." He tapped the side of his nose.

"Hey, these guys are the experts," Robert said. "Who are we to judge? They do financings all the time."

"True," said Kevin. "But it isn't necessarily in our best interest to overprice. Remember what Vinod said."

I shrugged. "What have we got to lose? I'll call Manzi, Gates, and Esber first thing in the morning." These were the CEOs of the Big Three software companies. Within a few days, I had appointments set up to meet with Manzi in Boston and Esber in L.A. But Gates wanted to come visit us personally to see the demo.

Robert and I took to the road. On the strength of my prior relationship with Lotus, Jim Manzi, its chief executive, assembled the bulk of his senior technical staff for our presentation. They were always alert for an opportunity to play in a new sandbox, and this sure looked like a big one. Manzi was all too aware of this tendency on the part of his staff, and so was understandably skeptical. He promised to get back to us after caucusing with his team.

In contrast, our meeting with Ed Esber of Ashton-Tate went quite smoothly. When we finished our pitch, Esber was visibly impressed. "I never realized I was a PC bigot," he said. "This really is different. We need to figure out how to get with it."

The next step was to host a visit from Bill Gates of Microsoft. He showed up promptly with an associate in tow. They were both dressed informally: slacks and white button-down shirts, open at the neck. Gates's sandy hair dusted the back of his collar and kept falling over his glasses. They looked like prep school boys exploring the city after college orientation. Although they had just walked in, Gates immediately slouched in front of a new Macintosh that sat unattended on the receptionist's desk, to see what software was loaded.

Robert cornered me before we went out to greet them. "Did he sign a non-disclosure? How much can we tell him?"

"I cleared the NDA with his attorneys, and he's supposed to be bringing a signed copy. It says in no uncertain terms that what we tell him is to be used only in considering a business relationship with us. He can't use what he learns for other purposes. Let's explain what we're doing, show him the demo, and pitch him on being an ISV."

Despite my reassurances, Robert looked concerned. "He's not going to like the fact that we aren't using DOS."

More than in most companies, Microsoft's culture reflected the personality of its founder. Gates had visions of dominating the PC software industry, banishing all competitors to a meager existence on barren, unfamiliar territory. He took the quick success of companies like Lotus as a personal challenge.

By any measure Gates was brilliant, yet he had never graduated from college nor accumulated the advanced degrees and academic accolades that he so richly deserved. Perhaps because of this lack of formal recognition, he seemed to have an itch that no single victory could scratch. Gates always seemed to be trying to prove that he was smarter than everyone else, and this competitiveness pervaded the entire company.

A shy technocrat with a squeaky voice, Gates found himself thrust into the public eye after the company's IPO in 1986. At first he did not take well to his success, fearing that it might soften his brutal work ethic and perpetual technical edge. As an improbable survival strategy, he seemed to cling to a view of himself as an underdog. Gates's newfound wealth, which was approaching the billion-dollar mark, had the unfortunate effect of making him a stereotype for the media, a money-culture symbol of the revenge of the nerds. He lived in a Spartan world of self-imposed austerity, flying coach class and parking his own car. As the wife of one of my colleagues once observed, "Bill is still trying to win the science fair."

Despite all this, I liked him. In most respects he seemed like a normal guy, albeit a bit awkward, with a knack for turning any discussion into a contest. I had mainly talked to him on social occasions, and found that it was a simple matter to put him at ease: defer to him on some point and he would open up. I respected his passion for work, his unmatched intelligence, and his genuine interest in mastering new ideas. When he wasn't feeling threatened, he was capable of discussing a wide range of topics, from politics to music, but he was most comfortable when talking about computer technology and gossiping about industry executives.

Gates seemed unusually relaxed that day during his visit to our offices. He greeted Robert and me with a broad smile and a firm handshake. I briefly wondered if he was more comfortable in business settings than social ones. He treated us as peers, avoiding the subtle social signals of higher rank. He helped himself to a can of soda from the refrigerator and followed us to the meeting room.

Robert described the concept of a pen computer, our approach, our product, and our desire for Microsoft to build applications for our system. Gates's eyes narrowed, and he leaned forward, folding his arms across his lap and staring at the table. At first I thought his stomach hurt. I looked at his associate for guidance, but he didn't appear the least bit concerned about Gates's condition.

"Go on," Gates said. As Robert continued, Gates started to rock back and forth like a *davening* Hasidic rabbi. This was his way of concentrating.

When Robert finished, Gates summarized the presentation in his own words, as though to ensure that he understood it completely, which he did. "This is something different, like the Mac," he said. "And that means we have a choice. We can start to work with you now or play catch-up if it succeeds."

Robert sighed, visibly relieved to hear this.

Gates then asked what application areas we would most want Microsoft to tackle. Robert made several suggestions, which he explained in some detail.

The two visitors spent a long time examining the demo, after which they had to rush to their next appointment. As we hurried to the door, Gates's competitive instincts, which had been conspicuously absent, reemerged. "You know, Kay Nishi and I talked about a machine like this some time ago." Nishi's company, ASCII Corporation, was Microsoft's marketing partner in Japan. "But the technology just wasn't ready. Maybe it's time now. I think you've got the right approach here."

"Thanks," I said. "I hope you'll work with us."

"When I get back to Seattle, I'll have someone call you to follow up," he said as the elevator doors snapped shut.

Several years later, Gates and I exchanged a series of letters regarding our recollections of this visit. He included a copy of an e-mail message he had sent to his senior staff following our meeting. Among other things, it said:

> Basically they are building a machine that Kay and I talked about building a long time ago — a machine with no keyboard and no disk . . . using a writing stylus with handwriting recognition for input . . . There are a few other people building things like this — in fact there was one discussed in the WSJ this week . . .
>
> . . . All the old ideas like using gestures for various commands they have "rediscovered" . . .
>
> ANALYSIS: This machine should be built as an open standard by a bunch of Japanese makers. The software layers should be more compatible with desktop stuff. Kaplan isn't the best CEO. They have some OK ideas but I don't think this thing will be big. We do need to think about note taking and the fact that small machines can be used everywhere especially with this input approach but I don't think we should be an ISV for them. [The full text of Gates's e-mail message is reprinted at his request in the Appendix.]

▣

Two weeks later I called Manzi, Esber, and Gates to find out if they were interested in investing.

"No, but this is an area we should know more about," Manzi said. "We'd like to continue discussions about doing applications."

"It'd present a conflict in working with other hardware companies," said Esber. "But I'd like to start a project with you."

"Not interested in the investment side," Gates said, "but I've assigned Jeff Raikes from our applications group to follow up. He'll contact you shortly."

Soon after these calls, Robert wandered into my office. "So where are we on the financing?"

"Nowhere. And we just killed a month to find out. Now we really have to get our asses in gear."

Kevin joined us, sporting his best "I told you so" look.

"OK guys," I said, "we have to split up the work. You two watch the shop. I'll put together a more formal business plan. I'll call John so we can start making appointments with the VCs. I'll need each of you to write up your areas, and then we have to put together a standard dog-and-pony show. Remember, these guys are mostly number crunchers, not visionaries. We need graphics, tables, what-ifs, and pictures. Let's review our budgets and figure out how much time we've got left."

Robert and Kevin went back to their offices and e-mailed me revised budgets, then rushed back in for the final count. Robert was a real spreadsheet jock, so he took over my computer, consolidating the various departments. "It looks pretty good. Seems like we can go until January if we watch it." January was six months away.

"That's a relief," Kevin said.

"Wait a minute," I said. "That doesn't sound right."

Robert reexamined some of the spreadsheet formulas. "Damn. These rows aren't included in the total."

I crouched at Robert's left side. "Oh shit!"

Now Kevin jostled his way in to read the tiny numbers on the screen. We looked as though we were playing a desperate game of musical chairs. "September," I said. "End of October at the latest. You guys have to give me more room than that. We'll get creamed."

They massaged the numbers until we could just make the mid-November payroll.

"OK," I said. "That's the plan."

When I got John Doerr on the phone, he started rattling off the names of venture investment firms for me to contact. "Merrill-Pickard, Sequoia, Sevin-Rosen, TVI, Accel, Mayfield, IVP, Oak —

they're all great venture groups." John was at his best when talking finance.

"Slow down, man, I can't type that fast!"

He paused only briefly. "No, wait. You can also contact our limiteds." He was referring to the organizations that put their money into Kleiner Perkins's investment funds as limited partners. He started right in again. "Chancellor Capital — that's Citibank; Aneas — they're the investment arm of Harvard University; General Motors Pension . . ." He was speaking so fast he could barely catch his breath.

"Hold it a minute, John." Every time I asked him to wait, he seemed to charge off in a new direction.

"We could also approach some corporate partners. Sun may be interested, Xerox has a venture group, Olivetti is looking for new technologies, Altos has some money to spare, I could call Andy Grove at Intel, Kyocera . . ." The problem was that each contact meant a series of dog-and-pony shows, partners' visits, and possibly one or more trips out of town. But the consequences of firing up more interest than we could handle wasn't John's top concern. ". . . and there are some investment banks we have relationships with that do venture stuff as well — Morgan Stanley, Warburg Pincus, Robertson Coleman, Hambrecht and Quist."

"John, I think that's enough."

He fell quiet for a second, as if to change gears. "Look, Jerry," he said, "the name of the game is to create a sense of scarcity. It's a game of numbers — the more people we interest, the more likely we are to get the money we need."

Another short silence, then he began to pour it on again. "There's a number of individuals you could cut in on the deal. The Sun guys — Bill Joy, Andy Bechtolsheim, Scott McNealy. Will Hearst from the *Examiner*. Sparky Sparks, he's a good guy. Steve Jobs. Ross Perot. Listen, I'll also send around some e-mail to the other KP partners to get their suggestions."

Now I knew I was in big trouble — this could turn into a freefor-all. "Enough!" I shouted. Then, softer, "I mean, I think I have enough to get started. Let's see what kind of response I get from these, and then if we need more, we can ask the other partners, all right?"

John seemed disappointed that I didn't want his partners' help as well. But he reluctantly agreed. "OK, OK. But they're ready any-time you want them."

I thanked him and hung up the phone. Calling in the KP part-ners can be like calling the fire department. They tend to show up in force and attack a project with benevolent but single-minded fury. The fire is sure to be out when they leave, but the furniture may be waterlogged and the windows broken. I figured we weren't that desperate yet.

"Christ, maybe we should install a revolving door," Robert whis-pered as he attempted to steer one visiting group into the demo lab while I led another one to a conference room in the back. Curious engineers would occasionally poke their heads above their cubicles to check out the latest crowd of visiting suits, like laying hens star-tled in their coops. Typically, we hosted two or three visiting dele-gations a day, and I would spend the brief interludes fielding phone calls, sending out business plans, checking references, and schedul-ing the obligatory follow-up meetings.

Most firms would first send in a senior partner and one or two associates. The associates would then return, accompanied by more-junior associates. These associates were the venture capital equiva-lent of fraternity pledges, brown-nosing their way through an in-ternship in the hope of someday making partner. Typically, they were recent M.B.A.s hungry for the next-larger-size BMW, eager to prove that they could make a startup's CEO jump through more hoops than the associates from rival firms.

Pierre Lamond, a general partner of Sequoia Capital, visited along with an associate.

"What's the pre-money valuation?" Lamond asked.

"Fourteen million." Each morning, I practiced saying this with a straight face in the mirror while I shaved. John and I had agreed to start at this valuation when we talked to the venture firms, knowing that they would be more price sensitive.

"We're on a short fuse here," I prodded.

"We can make a quick decision," Lamond's associate replied. Lamond nodded and then they left.

The associate soon returned with a research assistant, to study the market potential. The assistant set about applying the market estimation techniques he learned in business school, appropriate for introducing a new flavor of chewing gum into a crowded market but utterly irrelevant to our situation. "I'll call you and we'll go over these five-year market-share projections again," he said.

"Look," I said when he called, "rather than me reading this to you over the phone, why don't you just have your spreadsheet call my spreadsheet?" I was joking.

"Sounds good to me," he responded flatly. He continued to call us, requesting information for several weeks after his partners had decided not to invest.

Dave Marquardt of Technology Venture Investments dropped in to look us over. He asked to try the demo. Taking pen in hand, he signed his name on the screen. I looked at Robert nervously as the computer hesitated on the idiosyncratic scrawl. It translated as "MUD." Marquardt thought this was hilarious. I figured that was the end of it, but he continued to profess high interest in the deal. He was soon unavailable, however, away in Europe on an extended vacation when the time came for a final decision.

Arete Ventures, a firm specializing in utility industry financing, was inexplicably interested in GO. A small East Coast partnership, Arete was eager to coinvest with Kleiner Perkins. Jake Tarr, a young partner-to-be with choirboy looks and the boundless curiosity of a puppy, seemed to have an infinite amount of time to interrogate us. After he spent more than an hour on the phone with Kevin, going over the same subjects with him for the second time, Kevin begged me for some relief.

"Doesn't this guy have a life? What possible difference could it make to him how we plan to paint the plastic cases?"

"It could be worse — he could be totally uninterested," I said. But it was little comfort. "By the way, how *do* we plan to paint the cases?"

"Who the hell knows?" Kevin huffed, and stomped back to his office.

Jake Tarr's insatiable interest in the product was exceeded only by his curiosity about the other firms we were talking to. One day I called Lamond's associate from Sequoia, to take his temperature, and he mentioned that Tarr had called him to compare notes.

I was amazed. "How the hell did he know to get in touch with you?"

"I think he just called around town until he found someone you were talking to."

That shouldn't be too difficult, I thought. "What'd he say?"

"He wanted my opinion of the price," the associate said. "He thinks that fourteen mil pre-money is too high."

Tarr, it seemed, was already lobbying to lower the price. "And what do you think?" I asked.

"Don't know. We still have some more questions . . ."

"We add more than just money. We like to get involved with our companies." This was Jim Furneaux of Bessemer Partners. Since he worked for an out-of-town firm, he was selling harder than most. A large, soft-spoken man who looked as if he'd been drawn out of circles, Furneaux was based in Boston, although Bessemer was headquartered in New York and had a partner in Silicon Valley. "Our firm can move more quickly because we have a small, tight-knit partner group."

"That's great," I said. "It'd be a real help if we could limit the number of meetings and information requests. We're really getting buried."

"No problem."

On his first visit, Furneaux stayed for four hours. Then he talked to both John Doerr and Mitchell Kapor by phone. He called in a handwriting-recognition expert from Boston to evaluate our hand-writing technology. Then he asked for a list of references, all of whom he contacted, and a list of background reading material, which I assembled and sent. Next he met with Robert and Kevin again, separately. He even perused our "key man" insurance and company-mission statement. I was sure he was ready to make a decision, but this turned out to be just the beginning. "Now I have to get my partners involved."

He went home to Boston and asked his West Coast partner to come in. The man was skeptical; he'd been burned on a similar deal before. Furneaux quizzed Kevin yet again, this time about heat dissipation and display technology. He wanted some projections on handwriting-recognition speed. We cooked some up. Did we have any market research materials? I pulled something together for him. Then he asked another East Coast partner to visit. Furneaux returned to San Francisco and met with us again, as well as with Vinod

Khosla. But the West Coast partner was still unconvinced, so both John Doerr and Vinod Khosla drove out to his office to persuade him.

"My partners still have some doubts," the man said. So Furneaux called in another consultant to look at the deal. To my surprise, I learned that there was a fourth partner in New York, apparently the most senior partner, who was yet to be satisfied.

"I think we need to get you out there to visit," Furneaux said. So far, this quick investigation had taken more than eight weeks.

Jay Hoag, of Chancellor Capital Management in New York, returned my call. He was at a conference at Silverado, a posh resort in the Napa Valley.

"I could stop by on my way to the airport," he offered.

We gave him the standard pitch and demo, and sent him off to his flight. He called about a week later from another meeting, at the Princess Hotel in Scottsdale, Arizona.

"Is there anyone else I can share due diligence with?" This was a unique approach to asking who else was interested in the deal. "Due diligence," a legal term, refers to the responsibility of the investors to look into the companies they invest in with adequate care.

In a stroke of luck, I ran into John Doerr later that day outside an expensive Italian restaurant called Prego's, on Union Street — the yuppieville of San Francisco. I asked if he could take a few minutes to review the financing with me. To my surprise, he walked over to the curb and sat down. I sat next to him and brought him up to date. "Sounds good, just keep going," he said. Passers-by were trying not to notice the two grown men hunkered down in the gutter like wayward drunks. I wondered what they would think if they knew we were discussing a multimillion-dollar financing. Then I asked John what to do about Jay Hoag's request. "Tell him to call Sequoia. They've done their homework and will share it with Jay even though they've turned us down." It was a brilliant suggestion.

When I called Hoag back, he was off vacationing on Nantucket for two weeks. When he returned, I tracked him down at an investors' convention at the Hyatt Hotel in Monterey. It was now early October, and I was starting to sweat. There was a lot of activity and professed interest but no real commitments, and, more important, no lead.

"Jay, we need a decision."

"I understand. Fourteen mil pre is hard to swallow. But I'll be back in my office next week and talk it around."

I called John Doerr. "Look, we have about four weeks of cash and then it's the Doggy Diner. No one buys the price."

"All right." He sighed. "Let's drop it to twelve and close it up. Put out the word."

I hit the phones in a flash. "We have lots of interest, but no definite lead investor yet. I need to know if you are in or out at twelve million pre-money. Call me by five P.M. on Monday." On John's advice, I tried to make it sound like a last chance. For the six or seven people I could reach, I tried to answer their unresolved questions. For the remaining fifteen, I left a message on voice mail.

By Monday at six, not a single investor had called. They could smell desperation upwind in a blizzard. I felt like a plate spinner in a vaudeville act whose props had just come crashing to the floor.

"Jerry, this is a really interesting journal article about cursive-handwriting recognition." Todd Agulnick was standing in my doorway, holding up some papers. "We should contact this guy." I was exhausted and demoralized, and felt like telling him to pack up his belongings and go home for good. Our budget had called for hiring an expert in recognition, which we needed desperately, though I doubted it would matter anymore. But it was important not to scare the team, particularly Todd.

I looked at the name on the paper, one of those unpronounceable foreign names with strange accents above the letters: Rasha Božinović, Ph.D. The only evidence of the author's address was his apparent affiliation with the State University of New York at Buffalo. On a lark, I called the head of the computer science department there, who told me that Božinović had been one of his students, but he had returned home to Yugoslavia. For some reason, I found this funny. I imagined him tending sheep on a rocky slope on the Adriatic Sea, thinking deep thoughts about computational linguistics. In reality, he was doing research in a government laboratory. To my surprise, his professor had a phone number I could call. This was turning into a Hitchcock thriller, and I was hooked. A quick calculation showed that it was late evening there, about ten o'clock. I decided to risk calling. A pleasant-sounding older woman answered with something other than "hello."

"I'm looking for Rasha — calling from California, USA. Is he there, by any chance?" I avoided slaughtering his last name, choosing instead to risk the offense of sounding overly familiar. My words echoed electronically around the world, decaying into a long, hissy silence.

"Hello?" It was a young male voice.

"Are you Rasha?"

"Yes?" He sounded guarded.

"My name is Jerry Kaplan. I'm calling from California. I read your paper on cursive-handwriting recognition, and frankly, I would like to interest you in a job at my company in San Francisco."

There was a prolonged pause. I considered that his taking a call from the United States might get him in trouble with the local authorities. "You mean like the Golden Gate Bridge and crooked Lombard Street?"

"Yep, that's the place." Somehow, I knew I had him.

We talked for about fifteen minutes, then I went for the close. "If you don't mind my asking, how much are you paid now?"

"The equivalent of about two hundred dollars." He stopped, then realized I wouldn't know what that meant. "A month," he added. I figured this was some kind of mistake.

"We're talking in the range of fifty to sixty thousand dollars a year, plus benefits."

I heard a loud crash, as though something had fallen off a table. I didn't wait for an answer. "Is it legal for you to fly here for an interview?"

Rasha laughed and said, "At that price, who cares?"

I was starting to like this guy.

I sent him a round-trip ticket by international courier, and a few days later he arrived for the interview. He was a tall, thoughtful fellow with a prominent Adam's apple, and when he spoke, his voice rose and fell in gentle, rolling waves and his head listed from side to side. The engineers, too, liked him right away. I talked it over with Robert and Kevin in my office.

"Look, Jerry," Kevin said. "How the hell can we hire this guy, knowing that we could be out of business by the time he gets here?"

He was right, of course. The three of us were scared to death that all the hard work we had put in was about to go up in smoke, and here I was, talking about recruiting a guy from the other end of the world. I could see that the stress was taking its toll on my partners, as well as on me.

I was so tired that my mind began to wander as I tried to formulate an answer. It struck me that this must be what it's like to have incurable cancer. You wake up every morning wondering if it's a dream. Then you remember: there's something horrible that is slowly killing you, and you're helpless to do anything about it. So you go on acting out a normal life as though it makes a difference. I leaned on the window and gazed at the sturdy brick warehouse across the street. I noticed a small concrete cornerstone with the year 1904 on it and realized that the building must have survived the devastating earthquake in 1906. Somehow this seemed comforting.

Finally I thought of something to say. "The way I see it, it's a moral challenge. If Rasha shows up and we're dead, I'd feel personally responsible. On the other hand, we have an obligation to act on the assumption that we'll pull through. I think we should plan for success, even if it seems unlikely. We've searched high and low to find someone with his skills, and I don't want to set us back anymore. Besides, what would we tell the staff?"

"It's not that bad," Robert said. "We can reserve enough money to send him home with the equivalent of six months' pay, easy."

"Good point, Robert. Hell, let's do it."

"Joyride to hell," Kevin said, shaking his head. "Let's do it."

"John, this is your Tuesday wakeup call." I was talking to his voice mail. "No one wants to invest. We're screwed. What should we do?"

He called me back around noon. In contrast to my gloomy tone, he was upbeat. "I just got off the phone with Jim Furneaux of Bessemer. He thinks it's worth one more shot at his partners. If you can convince them, he's ready to be the lead. They have a meeting in New York this Friday. You can present to them there. Good luck." I hastily made travel arrangements for this one last-ditch effort to keep the company alive.

The meeting took place at the top of an art deco office building in Rockefeller Center. The New York partner sat in a Chippendale wing chair; he looked like a young Winston Churchill in suspenders. He sized me up with a smug growl as the others filed in and took places around the table.

"So what is that, some new kind of laptop?" He pointed his unlit cigar at my plastic mockup. He appeared to know little or nothing about our project. The other partners in the room, who had been loud and intimidating when I met with them previously, sat quietly as I looked around for some guidance.

The whole scene seriously annoyed me. I had spent months dancing for them while they dangled before me the promise of an imminent decision, and now it seemed that they hadn't kept one of their own partners adequately informed. My instinct was to tell them to stuff it and walk over to the Carnegie Deli for a decent reuben sandwich.

But then I thought about the crew back at the shop. Most of them had no idea what was going on — how close we were to losing everything and what Herculean efforts it took to keep the company funded. They had simply signed on to a dream and were ready to do whatever was necessary to pursue that dream, like a latter-day children's crusade. Until that moment, I had always found strength in the knowledge that if I really wanted to, I could bag whatever I was doing and just go home. This belief enabled me to accomplish hard things. But it was different now. Despite my own desires, I felt a larger obligation to the team: the only decent thing was to do what was best for them, not for me. Perhaps if they understood what was happening, I might have felt differently. But it was precisely their ignorance of the situation that placed the burden squarely on me. I now understood what it meant to be an adult.

I launched into my talk. It was the performance of my life, eloquent and passionate yet pragmatic. But the New York partner looked unconvinced.

"Now let me get this straight. This thing doesn't use a keyboard?" He stared at the plastic mockup in silence for several seconds. "We have to run to the airport," he said abruptly.

That was his way of ending the meeting. As the partners filed

out, Furneaux promised to call me at home the following morning.
I caught an evening flight back to San Francisco.

When I awoke the next morning, the sheets were damp with
sweat. I carried the cordless phone into the bathroom as I brushed
my teeth, for fear of missing his call. At last the phone rang.

"Jerry?" It was Furneaux. "Thanks for coming to New York. You
gave an excellent presentation."

"Thank you." I wanted to get to the point, but didn't want to be
rude. "So, can I have your firm's commitment?"

He exhaled for what seemed like a full minute. Then he spoke.
"Your presentation has raised a number of questions that we believe
deserve some further study . . ."

Beyond that, I wasn't listening. I was thinking about how to
arrange an orderly shutdown of the company. Naturally, as soon as
I hung up I called John Doerr, to deliver the bad news. He took it
stoically.

Then I said, "John, we have just enough money to give everyone
two weeks' severance."

He ignored this. "My partners meeting goes until five on Monday. Can you be there with Robert and Kevin at that time?"

"Of course."

When the three of us arrived at Kleiner Perkins, a thick San Francisco fog enveloped the building. We waited in John's office, staring
silently out the windows. There is nothing quite like looking out a
picture window at fog — the experience cannot be captured on
canvas or film. Your eyes have no fixed point to focus on, and so the
tableau has no discernible depth. It is the essence of gazing into a
void.

John strode in without saying hello. He picked up his multiline
office phone and set it down with a sharp crack on the small coffee
table between us. Its internal bell objected with a single *ding*.

He turned to me. "What did Scott Sperling say?"

"Huh?" I was still enveloped in fog.

"Scott Sperling of Aneas."

"Oh. You mean the guy from Harvard. That was a long time ago.
I think he said he thought the price was too high."

"What price does he think *isn't* too high?" John had a way of tossing out pointed questions that made you think you should know the answer. Robert, Kevin, and I glanced at one another quizzically, but we weren't really looking for an answer. We were too busy wondering what John was up to.

"I'm not sure," I said.

John punched the speaker button on his phone and started dialing. "Let's call him and ask."

"John, it's after eight o'clock in Boston," I protested. "He's probably gone home."

John looked me in the eye, expressionless. On the speaker phone we heard ringing, then Sperling's wife answered. There were baby noises in the background.

"Hi, can I talk to Scott, please?"

"Just a minute. He's got the baby on his lap."

A pause. Then, "Hello?" It was Sperling.

"Scott, this is John Doerr. I've got the GO management team here in the room with me."

"Hi guys." Sperling didn't sound at all bothered about being called at home.

"Scott, we need to close up this financing and we don't have a lead. Where do you stand?"

"We did a lot of work on this and we like the concept. There's a big market if you can make it work, but we feel the deal is overpriced at twelve million."

"At what price would you be willing to lead?"

We could hear Sperling gently chiding the baby as it grabbed for the receiver.

"Eight million pre," he said.

I whipped my calculator out of its holster like a six-shooter.

"And how much would you be willing to commit at that level?" John asked.

"Up to two million."

John pressed the phone's mute button. It began to flash red, like a warning light on a game show counting down to the buzzer. "What is that per share?" he asked me.

I was ready for his question. "Looks like about seventy-five cents. A twenty-five percent markup over the last round."

On the other end, it sounded as if the baby was gaining possession of the phone and might slam down the receiver at any moment.

John turned to Robert and Kevin, whose faces were white with fear. "Are you guys willing to do it?"

They were too shocked to speak. They both knew that for months we had cultivated Sperling, along with everyone else in sight, to no avail. Now, in one phone call, John had him ready to commit. Robert came to his senses and nodded his head vigorously.

"OK with us," I said.

John pressed the mute button again. "Scott, it's a deal. Jerry'll call you in the morning to get the paperwork started."

"Great, talk to you then," Sperling said. The baby seemed to be winning the fight for the phone.

"Scott, thanks. You won't regret it," I said.

John pressed the disconnect button, turned toward us, and said, "Congratulations, gentlemen, you have your lead." Then he left the room to attend to his next emergency.

Robert and Kevin were still staring at the phone, as though the governor had called with a full pardon just before our execution. The color returned to their faces so fast it looked like they were blushing.

"All that time — *all that time* — and it was only a matter of dropping the price!" Robert rubbed his face with his hands as he spoke. "I don't know whether to laugh or cry."

"Tell you what," Kevin said. "You laugh, I'll cry."

"Now that's teamwork," I said, replacing the phone on John's desk.

The rest of the week was a perpetual game of Dialing for Dollars. We raised over $6 million — more than our target of $5 million — in a matter of days. Suddenly the VPs were enthusiastic about project-planning meetings. Hiring plans were fired back up. As a joke, for Halloween the entire staff came in dressed as me — blue jeans, tie, sweater vest, and graying hair. The office hummed with energy again.

Rasha Božinović arrived from Yugoslavia excited but penniless, with his worldly belongings packed in a small trunk. He needed a place to crash, and a friend of mine kindly consented to let him use the guest suite at her newly renovated Pacific Heights Victorian house. His eyes went wide when she showed him his private bathroom, gleaming with marble and gold.

Within days, he had discovered reruns of a TV show called *Lifestyles of the Rich and Famous,* hosted by Robin Leach, a portly, money-culture sycophant with an English accent. Rasha would wander around the office imitating Leach's accent, moaning and fawning over everything from the office dishwasher to Todd Agulnick's Mustang convertible.

We closed the financing just in time to make our mid-November payroll. Once we had a lead and a realistic price, just about everyone fell into line. Except Bessemer, which still couldn't make up its mind.

5

························

The Customer

EVERY INDUSTRY has a trade show, a conference, an awards ceremony that marks its pace. Hollywood has the Academy Awards, publishing has the Pulitzers, and the computer industry has COMDEX. Every November, the COMDEX crowd invades Las Vegas, the only city in the country with facilities large enough to host the conference. It is inadequate to say that the conference is held *in* Las Vegas; for an entire week, the conference *is* Las Vegas.

Every hotel room within a hundred miles is occupied by a nerdy engineer or a computer salesman who used to sell cars. For the show, daily newspapers and round-the-clock television stations spring into existence with an efficiency that would be the envy of any military commander. All available convention facilities, hotel lobbies, and exhibition halls are crammed with booths where company representatives employ ever more outrageous tactics to attract attention. Walking at a normal pace for eight hours a day, it is not possible to traverse the miles of aisles during the conference week. Enormous pools of humanity shuffle along in streams that grow into waves that broaden into floods. At certain hours, there are long waiting lines for food, transportation, and that most critical modern resource, telephones. COMDEX is the Woodstock of the computer industry.

To celebrate closing the financing, I had taken the entire weekend off. On Monday morning, I flew out to Las Vegas for the fall 1989 COMDEX.

After checking into my hotel — no small logistical feat — I made the strategic blunder of remaining in the main exhibit hall of the convention center until it closed for the evening. When a hidden array of loudspeakers announced the end of the business day, everyone in the building — a cavernous structure the size of several football fields — poured out toward a waiting fleet of buses. Before me was a choppy sea of heads at a near standstill. Off in the distance, through the main lobby and beyond the entrance, I could see an army of crowd-control personnel in bright orange vests fighting a losing battle to channel the throng onto the buses. "Just take any one that goes near your destination," an exasperated marshal called with his bullhorn. I didn't think I'd ever make it to the buses and began looking around me, desperate to find another way out. It was then that I saw John Doerr, not too far away, waiting to leave too. This seemed an incredible coincidence — but then John had a way of bucking the odds.

"Long time no see!" I shouted.

"We're trapped like rats." He peered over the crowd, then added, "Follow me." We snaked around the mountainous multistory exhibits toward the west wall of the hall. "I believe there's a door behind those curtains."

Sure enough, there was. The door led to a private parking lot full of air-conditioned trailers, which served as makeshift conference rooms for executives of the major exhibitors. However, the exit was protected by a uniformed security guard.

"IBM," John said authoritatively. "Our tags are in H-Three." The guard waved us through.

"H-Three?" I whispered as soon as we were out of earshot. John just shrugged. We jumped a fence, and were free.

"We've been so focused on the financing, we haven't talked about what's next," I said, struggling to keep up with his brisk pace. "I'd really rather not go through that mill again if we can avoid it. That was too close for comfort."

"What do you have in mind?"

"We expect to have a prototype working by June. Then we'll need to start raising money, because we'll be running out again around next December. I'd like to approach some potential customers that might also be interested in investing later next year."

"Sounds like a good approach," John said. "Are you going to the Dvorak party tomorrow night? We can talk about it there."

"Sure."

After dark, COMDEX really comes alive. Every night there are lavish parties, often in sports arenas with big-name entertainment, and plenty of free eats. Between these affairs and the platters of cold cuts in the courtesy suites of hotels, only the forsaken or the foolish ever pay for food. The Dvorak party, an annual event, was legendary. Hosted by the computer columnist and commentator John Dvorak, it was supposed to be a secret, but everyone knew about it. It was usually held in Will Hearst's suite at the Desert Inn.

The next night around nine, I knocked on the door of the suite. "Who is it?" said a voice from the other side. This was followed by uproarious laughter, and the door swung open. Although the hallway had been quiet, the spacious two-story living room was packed, as was the indoor balcony overhead. The jokester was Stewart Alsop, an influential industry analyst and at that time the publisher of a widely read biweekly newsletter. Stewart was a towering man with a pot belly, a jolly face, and a semicircle of curly brown hair. No matter how elegantly dressed, he had the remarkable talent of looking like Fred Flintstone.

"It's Jerry Kaplan!" he announced, sounding like a talk-show host. Playing along with the drama, everyone in the room burst into exaggerated applause. "Uh-oh," he said, "better hide the food!"

"Food? Where?" I held out my arms as though to steady myself for a sudden start, in the style of Ralph Kramden.

John Doerr caught up with me at the buffet table. In the interest of increased efficiency, both Stewart and I had dispensed with the nicety of using plates and forks. We were exploring the multicultural spread with the evenhandedness of ambidextrous diplomats.

John selected a slice of cucumber cut in the shape of a heart, garnished with pimiento and a twist of lemon. "I put in a call to Naomi Seligman."

"Who on earth is that?" For once, I wasn't in the mood to do business.

"Ah, Naomi!" Stewart raised his glass of champagne as though making a toast. In his frame of mind, I couldn't tell for sure whether he really knew her or not.

John wasn't fazed. He clicked Stewart's glass with his bottle of Calistoga water. "Naomi runs the Research Board. A super-secret organization of the top CIOs in the country." At Fortune 500 companies, CIOs — chief information officers — were in charge of computers and information systems.

Stewart was now giving a running commentary. "An audience with the queen."

"She can see you December first," John said.

"Great," I said.

"In New York."

"Jesus, not again!" I said. "Sometimes I think you secretly work for the airlines."

"Hey, keep running up those frequent flier miles," Stewart said, "and soon they'll send you your own personalized barf bag."

"How come it's always Florida in July and New York in December?"

John ignored my protest. "If she gives you the nod, you can hit the forty top decision makers in one shot. This is how Jobs established the Mac."

Stewart disagreed. "John, none of those guys bought the Mac."

"Well, they should have." John started eyeing the miniature seafood burritos.

Inspired by the champagne, I had an urge to personally thank our host. "By the way, Stewart, where's Dvorak?"

"He never shows up."

"But it's his party!"

"That's nothing," John said. "Will Hearst isn't even in Las Vegas."

▣

For me, traveling to New York City was like an immigrant going back to visit the old country. People there live in a dirty, dangerous environment, bound to the land by some misplaced sense of heritage or tradition, unaware that the key to a better life is simply to get the hell out of town. Yet certain things are singularly authentic to the city and cannot exist anywhere else. One of them is Naomi Seligman. I flew to New York two weeks later to meet with her, as planned.

The cabby drove slowly down East Sixty-first Street, past a row of brownstones housing the gentry of the Upper East Side. I strained to read the numbers on the tarnished brass plaques above the doorbells. This wasn't the sort of place to put an office.

"This looks like it," the cabby said, rolling to a stop in midblock.

I got out, climbed the stoop, and rang the bell, not knowing what to expect.

An unintelligible voice crackled over the intercom, its speaker choked with urban dust. I figured I'd just announce myself. "This is Jerry Kaplan. I'm here to see Naomi Seligman." The speaker barked back and the door buzzed open. I entered and walked up a narrow stairway. A woman dressed in a business suit appeared to greet me.

"Mrs. Seligman?"

"No, I'm her assistant. She's running late. She apologizes, and asks that you wait in her office." I felt as if I were talking to a U.N. interpreter. She ushered me up yet another flight of stairs and into a large, elegant room at the rear of the building. The back wall was mainly a series of tall, shuttered windows looking out on a barren tree straining upward for sunlight from its cement cell two stories below. In summer, the tree would no doubt appear green and prosperous, but at this time of year it looked like the frightened victim of a mugging, limbs raised and trembling. A massive rosewood table sat in the center of the cavernous room. I took a seat in one of four imposing armchairs that ringed the table and reviewed my presentation while an antique wall clock marked each second with a loud, stern click.

Suddenly the door flew open, vibrating when it struck its doorstop, and a woman entered.

"Sorry I'm so late. The traffic was a nightmare." She dropped her packages and attempted to gesture with her wrist as she removed her glove. "If you wait past Thanksgiving, you might as well be shopping at the Central Park Zoo." I didn't have to ask who this was. She struggled to shed layer after layer of clothing — fur hat, long coat with matching fur trim, scarf, leather boots. She stopped at a cardigan and somber brown skirt, and walked over to extend a limp hand in welcome. After checking for messages, she took a seat next to me at the table.

This was my first chance to get a good look at Naomi Seligman, who had been in constant motion since her arrival. She had that look of youthful beauty shrouded by years of heavy responsibility. Her voice was a bit raspy and her tone was tinged with cynicism. "I'm sorry, but I have a dinner appointment tonight, so I can't stay long. What have you got?"

If I hadn't seen her New York style a hundred times before

among my mother's friends, I might have been intimidated. But I knew that it was just her way of being friendly, and that the proper reaction was deference. Besides, it all felt vaguely familiar, as though I were visiting a long-lost relative.

"Mrs. Seligman —"

"Call me Naomi."

"Naomi, I'll try to be brief." I pulled out the booklet that contained my presentation. "We're building a new kind of computer that works with a pen. I'd like to start by explaining who is going to use this new computer, and why they would want to." I turned to my first slide. "This pie chart shows the number of deskbound versus mobile workers in the U.S. population." Though only a few seconds had passed, I could see that she had instantly become impatient. She picked up her copy of the presentation and started paging through it. I gently reached out and turned it back to the first page. "I really think the best way to approach this is to understand the context of what we're trying to accomplish."

"Look, let's skip the fluff and get to the substance," she said, pointing to the plastic mockup I had laid on the table.

I closed the booklet and placed it back in my briefcase. "I can see that you're way ahead of me. You don't need the whole dog-and-pony." Instead, I picked up the plastic mockup, relieved that she had focused her attention away from me and toward the model.

"Ernie needs to see this," she said.

"Who?"

"Ernie von Simson, my associate."

I pictured a deposed count of the Austro-Hungarian Empire with Naomi Seligman on his arm, and suppressed the urge to laugh.

She walked over to a side table and hit a button. "Ernie needs to see this."

"He's out," said a meek voice at the other end. "But I think I can track him down."

"Please do." She returned to her chair. "Do you know about the Research Board?" She didn't wait for an answer. "The Board is probably the lowest-profile oraganization in the computer industry, as you can probably tell from the office. We avoid all publicity. The members individually control the computer budgets for some of the largest businesses in the world. We have a limit of forty members, strictly the heavy hitters. We meet three times a year. If they miss two meetings, they're out; if they change jobs, they're out. The information is strictly for the use of the members, and they keep it that way, so you don't have to worry about telling them things. Most of the meetings are closed discussion sessions, but occasionally we invite executives from the outside to present. I'm not talking about salesmen, I'm talking about CEOs — like John Akers of IBM, Ken Olsen of Digital. Once in a great while, we let young entrepreneurs in to present their ideas. Steve Jobs of Apple, Danny Hillis of Thinking Machines." She paused, then said emphatically, "I'd like you to present at our next meeting, in March." She gave me the look of a prizefight promoter signing a newcomer for his first big bout.

"Thank you, Mrs. Seligman."

"Naomi."

"Naomi, I appreciate the opportunity very much."

"When you present, just skip the bull. These are very smart guys, they'll get it."

"No problem."

"And no sales pitches. They hate that."

"No sales pitches," I repeated, as though taking orders. "Got it."

"And remember, they are all required to read our briefing book before the meeting, so they'll already know a lot about you."

The door opened and a pleasant-faced man walked in. He had dark thinning hair, a modest gray suit, and the demeanor of an English valet.

"Ah, Ernie, good," she said. "Take a look at this. We should show this at the next meeting. I'm sorry, but I have to go." She thanked me for coming, and walked out.

After introducing myself, I dove right in with a tour of the mockup.

Von Simson looked uncomfortable. "Excuse me," he said, "but do you have any background materials that we might review first? It really helps me to understand the context of what you are trying to accomplish."

The savvy business traveler knows that the advantage of flying United out of San Francisco airport is that you can park on top of the short-term lot and take a motorized walkway to the Red Carpet

Club on the mezzanine, avoiding the mayhem of the levels below. The door to this private club is identified only by a small, discreet sign, so as not to incite an uprising among the hapless souls waiting in long lines at the public counters.

March 3, 1989, was one of those days when this knowledge made a difference. I was scheduled to take a noon flight out of SFO to the Orange County airport, a short drive from the Four Seasons Hotel in Newport Beach, where the Research Board was holding its meeting. Because of a sudden storm, air traffic was in disarray. Adding to the confusion, the Orange County field was in the final stages of a major expansion, so all cars were detoured to a muddy, overcrowded hangar with too few gates.

As I entered the Red Carpet lounge, the departure display was urgently flashing the numbers of canceled flights, including mine. The two United attendants in the lounge anxiously tussled with their computer terminals, attempting to reroute the worried travelers surrounding them. Sizing up the crowd, I headed for a phone booth and dialed my travel agent, who I knew had access to United's computers.

"Julie, I'm screwed," I said into the phone. "My flight's canceled, the airport is a disaster, and I've got to get to L.A. for an important presentation."

"Hang on," Julie said. I could hear computer keys clicking as we spoke. "Everything's booked, but there's one seat not yet claimed in first class at eleven thirty-five, a possible no-show . . . But there's a waiting list."

"Who's on it?"

"Let me check . . . Looks like three gate upgrades, one Premier, and one Premier Executive."

I checked my watch. It was exactly 11:24. "Great, we can beat that. Put me in as a full-fare standby for first. The seat should be released within a minute, unless it's on gate hold." I waited, watching the seconds on my watch. I could hear her taking short, choppy breaths as she scanned her screen.

"Got it!" she said. As I hurried past the gate, I looked around and wondered which of the disappointed faces I had bumped to the next flight — if he was that lucky.

Since I had caught an earlier flight, I arrived at the hotel in

Newport Beach a half hour before I was expected. This created a bit of a problem for the Board's gatekeepers, who parked me in a make-shift hallway purgatory with a specially bound copy of their briefing book, containing only the pages that pertained to my presentation. When my time arrived, the door opened and I was invited into the meeting room. They were just finishing up a break.

The scene that greeted me was a feminist's nightmare — equal opportunity had yet to make its presence felt this high up on the org chart. Every chair was occupied by a smiling middle-aged male with graying hair. They were all informally dressed, and from the banter I overheard were clearly comfortable with one another — discussing their golf games, politely inquiring about the family, or swapping corporate war stories. Ernie von Simson called the meeting to order and made a short non-introduction, referring the members instead to the write-up in the briefing book and their earlier discussions. I wanted to keep my talk on an informal level as well, so I began by asking about their experiences with laptops. It was a good bet they would have some horror stories.

"May I ask how many of you have tried to automate a sales force with laptops?" Several hands went up. "And how many of those projects were successful?" I guessed right: there were groans and thumbs-down around the table. "I suspect that they were too heavy, too hard to use, and most of all, your salespeople were uncomfortable whipping them out in front of important clients." Nods around the room. "Well, I'm here to suggest a possible alternative."

The rest of my talk was more like a discussion, punctuated with questions and comments. Little by little, the Board's interest grew, as each member began to think of ways a pen computer could be used in their business.

"In our company, a really big issue is delivery of up-to-date equipment-maintenance information to field personnel," said the general manager of Rockwell International's information systems group. "I could see how an inspector could load in the latest bulletins electronically and refer to them as he walked around an aircraft or whatever, making notes directly on the computer. Then we could transmit his notes back to a central database after hours."

"We have a different problem," said an executive vice president at Merrill Lynch in New York. "Our people on the floor of the stock exchange make trades, which they write down on little slips of paper called trading tickets that sometimes are illegible or get lost. These errors can cost us big bucks. The runners could enter the trades on a pen computer and check for mistakes right on the spot. At the end of the trading session, we could download this information into our mainframes."

"Or," I added, "you could transmit a record of the trade immediately, through a wireless link. If the other trader had a compatible device, you could reconcile any discrepancies right on the spot."

A senior executive of the Bank of America said, "We could put them at tellers' stations and let the customer sign right there for a lot of transactions. Our field people could load in information about a big customer before a site visit."

The soft-spoken VP of data processing for State Farm Mutual Automobile Insurance sat staring at my plastic mockup, deep in thought. As one of the quieter members of the group, his words seemed to carry more weight than those of his more outgoing colleagues. "This could scratch a very big itch for State Farm. Our twenty thousand claims handlers out in the field are constantly on their feet. Laptops are out of the question. They can't sit down and type while they're inspecting a car or a building for damage. So they carry clipboards, and we're constantly transcribing forms that are full of errors and omissions. With this device, in principle the adjuster can complete the entire transaction and hand the customer a check right there." His eyes were bright and wide.

I was in heaven. Not only was the meeting going well, but it was a thrill for me to hear about these applications. Until then, my faith that pen computers would prove useful for businesses was based on theoretical considerations — that they would allow people to harness the power of computing when standing up and walking around, or when face to face with others. But I was less clear about precisely what these mobile workers were going to do with the machines. Hearing about specific uses, from people with real knowledge of those needs, gave the whole concept a new life for me.

One by one, the executives took a turn explaining the relevance of pen computers to their company, like true believers testifying at a revival meeting. Each one seemed to be trying to top the previous speaker.

Although I was over my allotted time by fifteen minutes, this spirited discussion went on. Von Simson suggested that the members take a break and continue in private, politely inviting me to leave. The entire group thanked me in unison, like a Greek chorus.

On my return flight up the coast, I reflected on the day's events. I had gotten a rare glimpse of a group of powerful movers and shakers, getting together to discuss issues of great import. By my estimate, the combined operating budgets of the forty Research Board members might be as much as $5 billion a year, more than the gross national product of many moderate-sized countries. The raw economic power of the Board's individual members was awesome — it could drive the creation of whole new technologies and industries, mobilizing armies of researchers, technicians, and factory workers.

Now I understood what IBM executives meant when they boasted that their company was "customer driven": they certainly weren't talking about personal computer users, whom they ignored for the most part. Rather, they were talking about their need to please the members of the Research Board and others like them at any cost. As I glimpsed the sculpted cliffs of Big Sur through a break in the clouds, I wondered which Board members were destined to become our patron saints and what mysterious forces they might set in motion on our behalf. I had to wait several weeks to find out, which gave me an opportunity to focus on some internal matters.

▣

By mid-March of 1989 GO had about thirty people on the payroll, so we needed to move to more spacious quarters. We selected an office complex called Metro Center, twenty miles south of San Francisco, in Foster City. This spanking-clean town was a city planner's dream. It was built on recent landfill, with no history of family farms, no arbitrary rights-of-way, no landmarks to work around. As a result, everything was tidy and new, laid out like a theme park — but in serious danger of falling into the bay in an earthquake.

We leased the entire fourteenth floor of the highest office tower, which had a spectacular view in all directions. I selected a modest-sized corner office with a broad floor-to-ceiling window, framing a panorama of the bay and the San Mateo Bridge. As a bonus, families of California turkey vultures circled just outside my window in the afternoon, using the building's warm updrafts to gain sufficient altitude to glide back to their nests in the western hills.

Once we had settled in, I called the company together for a staff meeting to discuss milestones. The team leaders had just decided to push back the prototype milestone from June to July, and I needed to ensure that they understood the cost of such delays.

"OK, let's get started," I said. "How do you like our new offices?" The room broke out in cheers and applause. "Now we look like a real company. We have a nice reception area, new furniture, a real phone system, computers . . . But we're missing one thing. Do you know what it is?"

"A company masseuse!" Todd Agulnick yelled.

"No — revenues," I said. The room went silent. "We look like a company, but we are only a venture. Ventures have investors, while companies have revenues. Every month we delay a revenue stream, we have to sell off more equity to stay alive. If we delay too long, the price of the equity goes down. So we have to sell off even more, and then eventually no one wants to buy. And you know what that means?"

"Burger-flipping time," said Mike Ouye.

"That's right, Mike. And like they teach you in Catholic school, time is money."

"Which Catholic school did you go to?" Celeste called out.

"Our Lady of the Almighty Dollar," I shot back.

"Funny, you don't look Catholic!" she said.

I forged ahead. "But I bet you never knew just how much time equals how much money. We spend about half a million dollars a month right now, about half of it on hardware development. At our current rate, we have enough money to last through the end of the year. At our last valuation, that means that every month we delay, we have to sell nearly seven hundred thousand shares of stock to stay afloat. Think about that. That's an average of twenty thousand shares for every man, woman, and child among us."

"Mostly children, if you ask me," said John Zussman, looking around the room.

"That's a lot. And it comes right out of our pockets. That's why our schedules are so important. It's as if we jumped out that window with the parts we need to sew a parachute." Everyone looked outside when I pointed. "From one point of view, it's a glorious ride until we hit the ground. That's when our paychecks bounce. But our plan calls for deploying the parachute by the third floor, for a soft landing. The first thing we notice as we fall is that there isn't a thirteenth floor to the building, so we're already behind. Then we discover that it takes longer to sew than we thought. Next, we're calculating how many bones we can afford to break when we hit. So let's not allow any more delays." Todd, who was leaning against a window, shivered and moved to a chair.

Mike's hand went up. "Why don't we just split up the five mil we have left and run for Venezuela? By my calculation, it would come to over one hundred fifty thousand a head."

"Plus the reward I can get by turning you in," John Zussman added.

I went on. "So, in conclusion, I have a little show-and-tell for you. Remember when you were in grade school and they had a charity fundraising drive and they would put up a big thermometer and color it in with red paint as they approached their goal? Well, guess what I've got." There were groans all around as I pulled out a three-by-four-foot oaktag thermometer that the office manager had constructed. "But on our thermometer, we erase the red as the money runs out."

We mounted the thermometer in the kitchen, and each month when the budgets were completed, the office manager would go in and mark how much money we had left. I noticed that people studied it nervously whenever they stopped in for a soda.

▣

A few days later, I got some bad news from Mitchell Kapor. In order to reduce the potential for frivolous lawsuits, his attorneys had advised him to resign from all the boards he was on. "I'm really sorry about this," he said. "I'm as enthusiastic as ever about the project, and I really do want to remain involved."

"Of course, Mitchell, I understand. No hard feelings."

My next call was to John Doerr, to see whom he might suggest as a replacement.

"How about David Liddle?" John asked.

"The cofounder and CEO of Metaphor?"

"Yeah, I think you'd really like him, and he would add a lot of value."

As usual, John was right. I met with David at his office in Mountain View. He was a tall, balding man with a ready smile, a slight paunch, and a love of sailing. He had a way of making you feel special, like your problems really mattered to him. It wasn't an act; he really did care. He always had an encouraging word or some sage advice, but his most remarkable characteristic was his ability to illuminate a discussion with a metaphor of dazzling brilliance. In my experience, David was living proof of two important things: you can be honest and caring and still get ahead, and there is always someone who is smarter than you are. From our first meeting, he treated me like a close friend.

"Jerry, if you're going to work with the kind of companies represented on the Research Board, you'll need a big corporate partner to back you up. When push comes to shove, these customers aren't going to buy hardware from a startup, at least not if they think they might become dependent on you. Their projects are going to require a level of support that you simply can't provide with your resources."

He saw the skeptical look on my face, so he called on his special skill. "When you dance with a bear, you can't slow down if you get tired."

"But how can I interest a partner?" I knew that Metaphor had such a relationship with IBM, and wanted advice on how to set one up.

"You won't have to do anything. The customer is your enticement."

Within a few weeks, David Liddle agreed to join our board of directors.

When I returned to the office, the receptionist told me that I had missed a call that sounded important. "It's from a Norm Vincent."

"The data processing VP from State Farm?"

"Don't know."

"What's the area code?"

"Three-oh-nine."

"That's it, central Illinois!"

I took the rest of the number and dialed Vincent immediately. Luckily, he was in the office. "Jerry, I really liked your presentation to the Research Board, and we're interested in following up. I'd like you to come out and meet with us, if you're willing."

Now I knew who had taken the the keenest interest after I left the Research Board. I was ecstatic.

6

The Proposal

RESEARCHERS who survey such things say that people who live within sight of mountains report being happier, on average, than people who don't. But whoever made that survey must never have visited Bloomington, Illinois. The easygoing residents of this city are as level-headed and homogeneous as the corn fields that stretch to the horizon in all directions. Through some invisible purity of character, or so it seemed to me, they have managed to tame the relentless forces of social and technological change that have dominated people's lives in the rest of the world. They share a sense of place and a pace so natural that it seems to spring up from the soil. Work starts at eight and ends at four-fifteen. Saturday is golf and a movie, and Sunday is church. Life appears so orderly that I suspect no one dies in Bloomington without his own consent.

As I flew south from Chicago one day in late March, the farmland looked like a giant checkerboard through the vibrating window of the prop plane. In the absence of natural features to divide the terrain, the early surveyors had simply laid out the roads in perfectly straight lines, a mile or so apart. The flight to Bloomington took only forty minutes, and I was tempted to pass the time by counting the number of squares we traversed as an estimate of the distance.

The fruits of local industry were proudly displayed in the tidy, small airport, including an automobile made at the Diamond Star Motors plant. There was even a bookcase full of donated paperbacks in the waiting room.

In my many years making business trips, this was the first time my host ever met me at the airport. Norm Vincent was accompanied by his wife, Donna, a practical and even-tempered woman thirteen years his junior. Her short brown hair and trim figure suggested an interest in sports. As with everything else in town, they seemed the perfect pair: he looked like a model for Grecian Formula, she looked like an athlete on a Wheaties box.

They took me out to dinner at a local bar called the Lucca Grill. "The usual?" the waitress asked.

"Of course," Norm responded. She brought Donna a 7-Up, and a beer for Norm. He explained that they come here every Thursday night. I was flattered to be included in their routine.

When our pizza arrived, Norm told me a little about State Farm. "The company was started by a farmer in a nearby town in 1922, but when his insurance business grew, he decided to pursue it full time." As he spoke, Donna stared at him with wide, adoring eyes, as though she were hearing all of this for the first time. "Eventually, he gave up the farm and moved to the big city."

I wondered what city he might mean. "Chicago?" I asked.

Donna pursed her lips at my foolishness. "No — Bloomington!"

"Since then, the company has just continued to expand like crazy," Norm said. "You may not realize just how large State Farm is. We're nearly twenty percent of the auto insurance business — that's about thirty million policies — and we insure about fifteen million homes. Handling all those policies, not to mention claims, takes an awful lot of data processing. This year alone we upgraded over fifty of our IBM mainframes."

This was an astounding amount of computing power, and a single mainframe can cost several million dollars. The infrastructure to support these giants — in terms of peripherals, power, and people — was hard to imagine. And he was only talking about their mainframe equipment, not the midrange computers in agents' offices. "Surely you're one of IBM's largest customers, then," I said.

"IBM doesn't tell us, but I think we're among their top ten commercial customers. Our mainframes and agents' computers are all blue." IBM is widely known as Big Blue, after its corporate color. "But in the late seventies, when it came time to put computers in claims offices, we decided to risk going with Hewlett Packard." To him, it seemed, this fifty-year-old, $12 billion company was a young upstart. I wondered why on earth he would consider dealing with GO.

After dinner, the Vincents dropped me off at my hotel, Jumer's Chateau. As I stepped from their new Dodge minivan, I couldn't believe my eyes. Sitting in the middle of a corn field was what appeared to be a European castle of questionable ancestry. Turrets topped a red-brick façade; medieval pennants flapped in the breeze; the doormen were dressed as royal pages. The lobby offered no reprieve: it was filled with a hodgepodge of imported artifacts and counterfeit antiques. A confessional functioned as a phone booth; a brass elevator car from France served as a planter. In the absence of a local historical tradition of any significance, Jumer's had accepted the challenge of bringing a touch of class to plain old Bloomington, and for only $65 a night.

The next morning, Norm picked me up at the hotel at eight A.M. for the short ride over to State Farm headquarters. It was a square complex of four-story buildings with a twelve-story tower near the center, planted in a sea of cars. Inside, a large atrium served as a sanctuary against the extremes of winter cold and summer heat. The place looked like a 1950s suburban high school, with flags, little reminders and aphorisms pinned to corkboards, and a large sign painted with "State Farm Welcomes . . ." followed by a space for the names of important visitors. We rode escalators up to the fourth floor, where Norm walked me through a maze of unoccupied space to get to his office. "This area is vacant because we don't have enough parking around the building for everyone," he explained. "We're working to correct the problem." It seemed that competing for spaces in an overcrowded parking lot had caused more discord than State Farm employees were expected to bear.

At precisely eight-thirty, we went into a nondescript conference room next to his office. About a dozen people were waiting, midlevel data-processing managers in Norm's group. Norm introduced me, and I began my presentation.

Yet again I went through my pitch, explaining the technology, the opportunity, and GO. But something was wrong. There was no glimmer of excitement, no moment of epiphany. They simply sat there politely, listening to what I was saying as though it were an algebra class. The higher I turned up the volume, the chillier they seemed to get. Finally I realized what the problem might be: here I was, a guy in a fancy suit from California, with a Jewish name, talking passionately about a crazy new idea in a place where conformity is prized. I toned it down and completed my presentation in as deadpan a manner as I could manage. This seemed to warm them up a bit.

At noontime, Norm took me to an executive dining room with two of his most senior people. The room looked like a coffee shop at a Holiday Inn: round laminated-wood tables, woven plastic placemats, and amber cafeteria-style glasses with ribs to ensure a solid grip. Across from my seat, a grinning multicolored cardboard Easter bunny was attached to the wall with double-sided tape.

When it was his turn to order, Norm asked for an iced tea. Everyone at the table stopped cold and stared. "In twenty years you've never ordered an iced tea!" the waitress said. "It must be spring if Norm is cutting loose like that," remarked one of his colleagues.

As we chatted, I noticed that the grinning bunny was starting to tilt, ever so slowly. I was so fascinated with its silent movement that I had difficulty concentrating on the conversation. It continued this graceful maneuver at an imperceptible pace throughout the meal, so that by dessert it was nearly horizontal. Finally, when the bunny could linger no longer, it fluttered to the ground like an errant leaf in a lazy summer breeze. The waitress quickly snatched it up and stuck it back in its proper place.

After lunch, Norm took me back to the airport. I could no longer contain my curiosity. "So, Norm, what do you really want to see happen here?"

"I sincerely believe you're on to something important, and it can make a real difference to State Farm. So here's the plan. We're going to put out an RFP for the auto claims estimating application." RFP is contractual lingo for a Request for Proposal, which usually contains the written specifications of a complex system that suppliers can then bid on.

"May I ask who else you intend to circulate the RFP to?"

"We're only sending it to four companies. IBM, HP, Wang, and GO."

I was stunned. The other bidders were thousands of times our size and had battalions of people who did nothing but respond to

RFPs. IBM would no doubt be pushing their own home-grown solution. HP was rumored to have some handwriting-recognition work in their labs and was already at war with IBM for the State Farm account. Wang was a surprise: this ailing Massachusetts company had demonstrated a simple pen-based attachment to a personal computer, but it didn't seem well suited for Norm's application. The question in my mind was whether GO would be taken seriously in this process or whether we had just awakened State Farm's interest, only to have the business stolen by one of its regular suppliers. The only way to find out was to give it a try.

"OK, we'll be happy to bid. Send us the RFP as soon as it's ready." We picked a date in late June for me to return and present our proposal.

As I came to understand over the next few years, Norm led a double life. He was a man whose intelligence and insight exceeded those of his colleagues. He was highly educated, widely traveled, and secretly passionate about technology. But to survive and prosper in Bloomington, he had learned to adopt a veneer of conformity. And his influence stretched far beyond Bloomington, as I was about to learn.

In the real world, management is an art, not a science. The formal techniques for decision analysis that business schools teach are fool's gold, a vain and misguided attempt to systematize the chaotic. The mere existence of these methods betrays a darker truth: we harbor a desperate desire to believe that the world is ultimately predictable. But anyone who has managed a startup knows that predictability is an illusion. In this environment, you are faced with an endless stream of arbitrary challenges that bear down on you with the relentlessness of an automatic pitching machine. There's no time to think. The trick is to know when to swing and when to duck. But I suspect that each arbitrary decision is like smoking a pack of cigarettes — somehow, somewhere, it shortens your life.

After many months of phone calls and discussions, we were at last ready to start a relationship with Microsoft. After the meeting with Gates, Jeff Raikes — a VP in Microsoft's applications group — had called to begin exploring the possibility of building applications for our system. We negotiated a cooperation agreement in which he was to supply people who would study our technology and evaluate the opportunity for Microsoft, and we would train them and provide technical support.

Predictably, the sticking point had concerned protection of our intellectual property. The final language read: "Each party agrees . . . to use [the other party's confidential] Information only for the purpose of furthering this joint project . . . The participation of Microsoft and GO staff in joint design and implementation efforts will not create an interest or ownership on behalf of either party in the other's proprietary, confidential, or trade secret information." In order to reassure me on this point, Raikes reiterated a policy that I was familiar with from my Lotus days, that Microsoft would maintain a "Chinese wall" between its two divisions, so I needn't worry about GO's confidential information jumping from the applications group to the operating systems group — which we saw as a potential competitor. The agreement called for Microsoft to assign personnel and for us to provide them with "temporary office space, access to proprietary design documentation and related information, and technical cooperation." Finally, we got a draft with Raikes's signature for us to countersign.

Since this was really a software cooperation agreement, Robert was to sign it. He checked with me one last time. "OK now, you're comfortable that we can go ahead?"

"I don't know that 'comfortable' is the right word," I said. "This is a calculated risk. But I don't think we'll get any further information by waiting that will sway us one way or the other." I reread the signature copies, just to be sure. "The language seems quite clear to me. I don't see how they can steal our stuff without blatantly violating the agreement."

"OK then, I'll execute it and we'll get under way," Robert said. He signed both copies.

A few weeks later, Celeste Baranski returned from lunch looking pale and worried. She had just learned from a friend at another startup that Yamaha's electronics division had changed the specs on its EGA controller chip, which manages the display of information on computer screens.

"So?" I asked.

She looked at me as if I were some kind of a dolt. "We're using the same controller chip that they are!"

Kevin dropped his pen. "Shit, and we just gave the go-ahead for fabricating our prototypes. Celeste, you better get the changes in process ASAP."

Later, I talked to Kevin privately. "A few weeks ago we had to beg to get some experimental SRAMs from the Japanese. Now Yamaha screws us on the EGA controller. What the hell's going on?"

He was unperturbed. "That isn't the worst of it. Our electronic pen and sensor supplier in Tokyo isn't being responsive either. We don't have the parts they promised us, and they've been vague about the specs." In contrast to SRAMs and EGA chips, for which we had other suppliers if worse came to worst, the pen and sensor were nearly unique, and these critical parts were central to our hardware design.

I was furious. "Why are our suppliers trying to put us out of business?"

"The problem is face," Kevin said.

"What?"

"Actually, lack of face. In Japan, if you're out of sight, you're out of mind. They just don't think the way we do here. We aren't real if we aren't there."

"But you've been going over there regularly!"

"True. But *you* haven't, and you're the CEO. In Japan, hierarchy is everything."

I thought for a moment, then said, "Let's go."

Within a week, we were on a flight to Tokyo on Japan Air Lines. The seats were not designed for the comfort of Westerners, so my knees pressed against the seatback in front of me. The passengers were mostly returning Japanese businessmen, which meant that they all smoked. Having been rudely deprived by San Francisco's draconian no-smoking laws, they all lit up as soon as we were safely airborne.

Eleven hours later we pried ourselves out of our seats and, after waiting an hour, piled into an equally cramped bus bound for our hotel in downtown Tokyo. Taking a cab was out of the question — it would have cost upward of $150, since the airport is nowhere near the city. Under normal conditions, the trip would have taken an hour and a half, but it was rush hour, and drizzling, so it took an extra hour. When we approached the Tokyo Hilton, a loud chime warned us of our impending arrival, and a relaxed and refreshed recorded female voice cheerfully announced the destination in both Japanese and English, a stark contrast to the mood of the passengers.

The next morning, a bright yellow sun burned through the urban smog. Our liaison from the pen supplier met us at the hotel with a car and driver for the hour-long ride to company headquarters. This common Japanese practice is not just a courtesy, it is a practical response to the difficulty most foreigners have in getting around. In many places, street addresses are numbered not in physical order but according to when the structures were built.

Arriving at the company's building, we exchanged our street shoes for slippers. The offices bustled with young, enthusiastic engineers, all wearing identical gray smocks over their clothes. We met the team assigned to our project in a conference room. They promptly set to work, poring over Kevin's diagrams as though they had just unearthed a lost religious text. In the course of the day, we discovered the reason for their intense interest: they knew of a problem with our design that they had been too embarrassed to point out via fax or phone, presumably for fear of offending us. I waited patiently while the group worked out a solution. The process took several hours, so I had plenty of time to look around.

While the room appeared plain and orderly, there was one thing that seemed out of place: mounted high on a wall was a photograph of a man, a woman, and a boy. During a break, I noticed the same picture in the next room, and the next. I pointed this out to Kevin when we were momentarily left alone in the conference room. "Who do you suppose that is?" I asked as quietly as I could, pointing to the picture.

Kevin turned white. "Oh my God."

"What is it? Are you OK?"

"It's the Reverend Sun Myung Moon and his family."

At first I thought he was joking, but then I took a closer look.

"Holy shit, they're Moonies! Christ, Kevin, don't tell me our project is dependent on people who pass out flowers in airports."

"Look, all I know is that these guys are great engineers. Their pen technology is the best I've seen."

"Yes, but for all we know, they might require us to include leaflets with our products."

He just shrugged.

In the middle of the rush to get ready for State Farm's RFP, we hosted our first visit from Microsoft. Jeff Raikes had assigned two people to work with us: an inquisitive young woman from his marketing department and a talented young engineer named Lloyd Frink, who looked and acted like a junior Bill Gates. I instructed our staff to provide them with the marketing information they would need to evaluate opportunities for applications and with the technical information required to do development — but no more. Specifically, they didn't need to know about State Farm.

Frink, the engineer, took up hours of our time, debating our approach at every turn. He was enamored of Windows, and couldn't imagine why it wasn't suitable for our application. "I thought he came here to build applications, not to convince us that we're wrong," Robert confided during a break.

"Look, why don't you walk him through our logic one step at a time," I suggested. "He's a smart guy, he'll catch on."

Robert put together a detailed list of issues, hypothesizing a product called Windows/H, for Windows with Handwriting. He did a masterly job of explaining its shortcomings. Frink took notes, but hardly seemed convinced. He spent many hours studying our documentation to understand how we addressed each of these problems.

A few weeks later, Robert was invited to brief a larger group of people in the applications division at Microsoft headquarters, in Redmond, Washington. He was thoroughly disgusted when he returned. "All they did was beat me up about why we should be using Windows. They insinuated that if we didn't, they might do this themselves. After my presentation, they arranged a series of meetings to convince me, including several with key members of the Windows development team."

"I thought they were going to keep this stuff separate from the operating systems group," I said.

"Based on what I saw, I doubt that very much. I don't think they're serious about working with us."

"What if they actually *do* start a project based on Windows?" I was asking myself as much as him.

Robert looked frustrated with me. "It'll suck unless they want to make major changes, just like I've been telling them. It's just a plain fact that the requirements for a portable pen computer are very different than for a desktop computer, as you know." He didn't have to tell me this; it was the reason we were in business in the first place. But he was getting used to explaining this over and over, like a broken record.

"Look, Robert, we don't have time to sell uphill with them, we've got other fires to fight. Unless they follow up, let's let the matter drop at this point. We can make another pass at them when we're further along."

"That's fine with me." He scurried off to a meeting about the upcoming State Farm presentation.

Microsoft never called to arrange a follow-up visit, and neither did we.

As the date of our presentation to State Farm grew near, I was able to learn that IBM was scheduled to present the day after us. IBM's strategy was to do what they knew GO could not: lavish attention on the State Farm folks. A corporate jet would whisk them off on a tour of various IBM facilities, where they would be lectured by noted IBM researchers, witness demonstrations of advanced technologies, and experience the power of the sovereign state of IBM. The point would be made loud and clear: IBM is so large and powerful that it can do anything anyone else can do.

Our only hope was to show a distinct technological edge. The problem was that although we might be able to assemble a working unit by then, and even have some software to demonstrate, there was no realistic chance of putting the two together before the meeting. It was time to develop Plan B.

I called a meeting with the staff. "Look, it isn't essential that we show State Farm the software in its full glory. For one thing, we have no way to demo to a large crowd, and you can bet there'll be one. We've put together a paper presentation with lots of pretty images of car parts. All we need to do is give them an idea of what it will look like on the screen."

Phil Ydens, the quiet engineer, spoke up. "I think Mike and I can write a special program that will display one image. But you won't be able to do anything else with it."

"That ought to do it," I said. We picked out an exploded diagram of what we speculated was a drive train, but no one knew for sure — it had lots of wheels floating around a central shaft. Then I went on to the next issue. "The critical thing we really have to have is a working hardware unit. Otherwise, it's all acetates and yak-yak."

Kevin looked worried. "Everything would have to go right from now until then. Based on our experience, I wouldn't bet on it."

"But you already have bet on it!" Celeste said. "That's why I'm a nervous wreck."

She was referring to a challenge Kevin had accepted at the previous board meeting. She had spent the last several months working on a custom chip, known informally as a "glue" ASIC, that knits together all of the components of the system. Once her design was complete, she turned it over for fabrication, a process that would take about eight weeks. John Doerr, skeptical about our schedules, had doubted whether a custom chip this complex was likely to work the first time. If it had to be redesigned, we would be set back at least four months. In a show of confidence, and to demonstrate his faith in his team, Kevin had bet John $100 that Celeste's GOLD chip, as she had named it, would work on the first pass.

With less than two weeks to go until the State Farm presentation, the last of the hardware components came back from the manufacturer. We had enough parts on hand for five units, out of which we hoped to assemble two or three working ones.

Kevin and Celeste turned the hardware lab into a room that looked like a pediatric intensive care unit, with electronic monitoring devices surrounding special clean tables. Probes and wires clung to every quadrant of the precious, fragile circuit boards struggling for life. The whole staff was on call in case someone could provide the missing piece of information or skill that might save a dying unit. Anxious engineers would hover in the doorway, waiting for news, only to be shooed away. To reduce the confusion, Celeste banned everyone from the lab, but periodically sent one of her technicians out with progress reports.

"The CPU on unit three is booting," she would announce.

"Wow, a heartbeat!" yelled one of the engineers. Everyone applauded.

Every so often the technician would arrive at my office winded from the brief sprint. "We've lost three of the units, but Kevin wants me to tell you that we still have enough parts for two more, if we cannibalize those," she reported.

After the third day, Kevin called an emergency meeting with me and Robert. He looked like hell. "OK, here's the scoop. We have two units where everything looks fine except one thing."

"What's that?" Robert asked.

"They're both blind — no output on the display. I can tell by probing the board that the signals are there, and I can prove that the displays are working, so the problem has to be in the Yamaha EGA controller. I've faxed the Yamaha engineers in Japan several times, but they keep insisting that the chip works fine."

I exhaled thoughtfully. "Could it just be the ones we have?"

"Not likely. They pass all the diagnostics, and all of them behave the same way."

Robert looked doubtful. "How can their engineers insist that the chips work?"

"The problem is that the chips work fine in DOS machines, but the DOS machines don't exercise all the functionality of the chips," said Kevin. "Yamaha doesn't understand that we aren't just building PC clones, though I keep telling them that."

This was the first of many times we were to discover just how fragile computer technologies can be. If you stray from the mainstream, there is no guarantee that accepted, time-tested components will work as expected. It's a small miracle that computers are reliable at all: they function only as the result of endless trial and error, folklore, and superstition.

"What should we do?" I asked.

Kevin was ready with a battle plan. "I'm asking my staff to switch to a different schedule — Tokyo time — so we can interact directly with Yamaha during their working day, which starts around two P.M. our time and goes past midnight. Let's set up a room here where we can sleep, if necessary."

"Consider it done," I said.

"Then we're going in for saturation bombing — phone calls, faxes, demands that their engineers come here, whatever it takes — until they send me the schematics for their chip so I can debug it myself."

For the next several days, weary engineers labored around the clock, developing ammunition to support the hardware staff during the day and faxing volumes of test reports to Japan in the evening. The team was getting desperate as the deadline approached. They were ready to try anything. Finally Kevin got a schematic out of Yamaha, but only for part of the circuit.

I entered the lab, violating Celeste's ban. She was sitting there exhausted and demoralized. Kevin was calm, studying the diagrams yet again. "It's not going to work," Celeste said, ignoring my lapse of protocol. "We've tried everything. Their goddamn chip just doesn't meet the spec. See if I ever use their parts again!"

Kevin silently ignored us. Then he spoke. "Celeste, look at this." He pointed to the diagram. "This control line is drawn differently from the rest."

"So what? They can draw lines any way they damn well please."

"Well, maybe it *is* different," Kevin said. "Maybe we need to hold the signal high for longer than on the other lines."

"You're nuts!"

"No, really."

Celeste threw up her hands in resignation. "Well, I'm willing to try anything at this point."

I returned to my office, thinking about how the State Farm presentation would come off without a working prototype. A few minutes later, Celeste and Kevin walked in. Kevin was calm as always, but Celeste appeared ready to explode. They stood there without saying a word. Celeste looked at Kevin, but he deferred to her, gesturing for her to speak.

"I can't believe it. I can't friggin' believe it." Kevin started laughing as she spoke. By now a small crowd had started to gather. "I simply can't believe it." She repeated this over and over.

"OK, OK, so what happened?" I asked her.

"It worked. Kevin's crazy idea actually worked. I just can't believe it. When I designed the GOLD chip, I had a spare flip-flop, so I wired it up to three extra pins, which I labeled EFUFD, EFUFQ, and EFUFR."

"What does all that stand for?"

"Emergency fuck-up flip-flop pins D, Q, and R. I never thought we'd need it, but I used it to change the propagation delay on one of the output lines, like Kevin suggested."

One by one, we filed into the hardware lab to see the newborn. The people waiting their turn in line outside could hear oohs and aahs. There on the table, with its insides splayed out for everyone to see, was a pile of electronics with a picture of a drive train on its screen.

There was only one cherished working unit, and now the team undertook the delicate task of assembling it into the slim, inch-thick black case. Robert took it to his office to practice for the State Farm presentation while Kevin and Celeste went out for a quick lunch.

Stepping off the elevator when they returned, Celeste suddenly looked up and down the hallway, her eyes filled with fear. "Kevin, do you smell something?"

He stopped and sniffed. There was the unmistakable odor of burning circuitry in the air. They raced down the hall to Robert's office, where they found him staring at his desk, looking as if he had just run over a small child. In front of him was the precious prototype, scarred black with smoke, obviously ruined.

"Software people!" Celeste said in disgust.

"I was just writing on it when flames shot out through the connector slots," Robert said.

Now Celeste was truly ready to give up. But Kevin was unflappable. "Maybe it's just a hairline short on the circuit board," he said. "We've got enough parts left for one more."

They returned to the lab and assembled the final unit, and miraculously it worked. Like proud parents, they brought it to me swathed in soft towels. It looked exactly like the plastic mockup I had been touting for the past nine months. I pressed the on button and the screen came to life. It was pure magic — no one had ever before seen this much computing power packed into such a diminutive, four-pound frame. Someone snapped a Polaroid. I wrote on it: "6/20/89 — First Working Unit."

Celeste demonstrated how to change the battery. "Now be careful. The battery should last about an hour, so don't play with it until your presentation. Here's a spare in case it drains while you're on your trip."

"Not to mention that there's no guarantee how many times it'll come on, so I suggest you don't wake it up unless you have to," Kevin added. "While you're away, we'll try to bring up a second unit."

It had come right down to the wire. We were scheduled to depart for Bloomington the next day.

"These puddle jumpers make me nervous," Robert said as the pilot circled the airport. After landing, we loaded our gear into a taxi for the short ride to Jumer's Chateau. Although it was only late June, it seemed that in Bloomington every summer day was the Fourth of July. I pointed out to Robert a mint-condition 1975 red Corvette — I could tell the year from the license plate. I noticed red, white, and blue bunting on a number of buildings, in anticipation of the upcoming holiday. Then a pair of shiny silver Corvettes passed us by. At the next corner, four of the eight cars waiting at the light were Corvettes. Everywhere we looked there were Corvettes of every imaginable color and style.

"What the hell's going on?" I asked the cab driver.

"Huh?"

"We're surrounded by Corvettes!"

"Of course. Isn't that why you're here?"

Robert leaned over to me and hummed the theme from *The Twilight Zone.*

I was getting worried. "Hell no. We're here for a meeting at State Farm."

"Oh." He looked at us in his rear-view mirror. "I thought you were here for the Corvette Corral."

The cab driver told us that each June, every Corvette owner within a thousand miles brings his car to Bloomington for this event.

"What do they do?"

"Just hang around, talk about their cars, drink beer, swap parts — you know."

I didn't know. I owned a ten-year-old yellow Mazda station wagon, and I couldn't imagine going to a Mazda station wagon convention.

The next morning, before our presentation, Robert and I had breakfast at a restaurant called Bob Evans, a bright red wooden structure with white trim, designed to look like the television ideal of a midwestern barn. The inside was decorated with lanyards and bales of hay. The place reeked of cholesterol, and judging from the clientele, they had obviously been eating breakfast there for years.

Robert bit into a wedge of toast dripping with butter. "I think I can hear my arteries hardening." Not wanting to buck the theme, we ate eggs and bacon while reviewing our presentation one last time.

"Do you think we should check the prototype?" he asked.

"Let's not risk it. Kevin warned me that it could die at any time. Let's just leave it in the hands of God."

"I'm glad to see you're coping with all this strain by turning to religion."

We were both used to performing under pressure, but this presentation was particularly important. We knew that if we lost the bid to IBM or Hewlett Packard, we were in trouble — back on the fundraising trail with slim hopes for success, given how hard it was to round up investors last time.

I took the bait. "Please join me in prayer." I cupped my hands. "Dear Lord, though we walk through the valley of dead batteries, deliver us from EGA, and give us this day our daily bid."

"Amen," Robert added. The elderly couple at the next table looked on approvingly. They apparently thought we were serious.

By the time we entered the fourth-floor conference room at State Farm, we were both wired on caffeine. Neither of us was in the habit of drinking coffee, but our presentation was scheduled to start at six A.M. San Francisco time, and we wanted to be on our toes. I laid out the overhead slides, handouts, and copies of our response to the RFP in neat piles on the table alongside the prototype, which I placed in plain view. Robert was dismayed — he was hoping for some higher drama. "Don't worry," I whispered to him, "they'll think it's the plastic mockup. I've got a plan."

"That's what Nixon said about ending the Vietnam War," he whispered back.

Soon the seats began filling up. I could see that just about everyone who had been at either of the two previous meetings was present. Norm Vincent entered the room last, greeted Robert and me, and introduced us. He took a seat at the head of the conference table, right next to the prototype, where the group could read his reactions.

I began by reviewing some of the same slides I had shown in the earlier meetings, covering the potential for pen computing, GO, and our product plans. I explained that while I knew most of them had seen these slides, I wanted to be sure we were all starting with the same background information. My real intent, though, was to lull them into a false sense that they knew what was coming next. At the point in my talk where I usually pass around the plastic mockup, I reached over and picked up the live prototype, which looked exactly the same. Several people nodded knowingly.

"Last week, Norm asked me if I'd have anything new to show today, and I told him no. Well, our engineers have been working around the clock to prove me wrong. You've seen this model before, but I have a surprise for you. Today, for the first time, I'd like to show you an actual, real live working pen computer."

The audience, which had been characteristically reserved and mannerly, snapped to attention as though someone had shaken their chairs. Everyone strained forward to get a closer look. I stood the unit upright on the table, paused dramatically, and glanced at Robert. He had the look of a defiant death-row prisoner about to give the signal for his own execution. He nodded sharply, then looked toward heaven. I pressed the on button.

A long vertical line flashed on the screen, but it unexpectedly went blank. Robert sighed. And then, without warning, the picture of the drive train appeared. Norm gasped. The audience, accustomed to years of boring presentations, spontaneously burst into applause. Most of them were from the world of mainframe data processing and had rarely experienced the impact of a graphical computer application, much less one on a lightweight, portable tablet. They knew they were witnessing history. "I want you to know it was only yesterday that this unit — the first pen computer — was assembled. You are the first people ever to see one working."

Robert and I appeared to be flushed with excitement, overcome by the significance of the event. In fact, our reaction was due at least as much to our relief that the unit hadn't burst into flames, heightened by the dose of caffeine rushing through our veins. I handed the prototype to Norm Vincent, who inspected it closely. As he passed it around the room, I turned things over to Robert, who gave a tour of the application we had designed to process their automobile claims.

As Robert explained, the application used our notebook metaphor. The claims estimator would turn to a page for the type of car being examined and tap on the quadrant of the car where the damage had occurred. The application would then display an exploded view of the car's relevant parts. By tapping with the pen on an individual part and checking a box to indicate whether it should be replaced or repaired, an automatic estimate of the job's cost would appear on another page in the notebook. If there was a cheaper alternative replacement, or if repairing a part would cost more than buying a new one, the computer would flash a warning. The completed report, possibly including a scanned "snapshot" image of the damaged vehicle, could be transmitted to the State Farm mainframes. The audience recognized that our approach was simple, elegant, and practical.

But there was still one important matter to attend to: heading off the competition. I needed some way to extend the emotional impact of our presentation through the following day, when this same group would be wined and dined by IBM, their familiar and trusted partner.

"Tomorrow you are going to visit IBM. They are flying you to New York on a corporate jet; they will feed you a wonderful lunch; they will show you some exciting demonstrations. Their work is very good, and given enough time, I'm sure they can do a creditable job on your application. However, throughout their show, I ask you to bear in mind just one fact: there is a difference between a dog and a cat.

"Now, State Farm has mice in the house, and we have designed a cat." I pointed to the prototype. "It is tuned precisely to meet your requirements. It's the right tool for the job. But IBM has several dogs they want to sell you." The audience started to giggle.

"They will raise the first curtain and show you that they can make a dog that looks like a cat. They shave off the hair, put it in a fur coat, and tack whiskers on it. Then they will close that curtain and open a second one, where they exhibit a dog that is trained to act like a cat. It eats cat food, and meows. Next, behind the third curtain, they will show you that they can fit a dog into a cat carrier. They will cut off the ears and the tail, amputate a leg, and sure enough it fits in the carrier. It can't run very well, but it fits. Then they will have a noted lecturer explain their plan for breeding a dog that chases mice." By now my audience was laughing out loud.

"Throughout this process, I ask only that you remember one thing: it is in their interest to sell you a dog, but it is in your interest to buy a cat." Norm led the group in a round of applause.

On the way to the airport, Robert smiled as he gazed out the taxi's window. "Dogs and cats." He chuckled. "Critter would have been proud. I just want you to know that that was great. That was really great." Coming from him, this was high praise indeed. Our respect for each other usually went unstated.

"Thanks," I said. "So were you."

The following Monday, I got a call from Norm Vincent.

"So," I asked, "how did the IBM meeting go?"

"It was OK, but they couldn't figure out why our people kept snickering. You really set the tone with that dog thing."

"I hope you didn't mind."

"Not at all," Norm said. "The presentation was excellent. You really set your cause ahead last week."

"Thanks, I'm glad you liked it."

"Jerry, I don't know how you're going to take this." There was an ominous pause. I braced myself to hear that he had nonetheless decided to work with IBM or HP. "But how would you feel about working for one of our regular suppliers?"

This was so far from my mind that at first I didn't understand what he meant. "You mean you want me to quit and go work for them?"

"No, I'd like GO to team up with them on the project — that is, if you're willing."

This was the best of all possible worlds. Not only would we get State Farm's business, but we could neutralize potential competitors by becoming its pen computer supplier. State Farm could open the door to a strategic partnership with a major computer company. With one big customer in the bag, we could likely get the partner's support on a wide variety of fronts — like approaching other customers, manufacturing our products, promoting applications development, and most important, funding GO. I was thrilled. But there was one missing detail.

"We'd be happy to, Norm. But which supplier do you have in mind?"

"I'll leave that completely up to you."

As if things weren't good enough already, he was indifferent about whom we worked with. I felt like a grateful teenager who had just been promised the use of a Rolls for the prom, only to be told that he could also take his pick of the town's most beautiful girls to escort.

"You mean I can choose either IBM or HP?"

"That's right."

Now that I was firmly in the driver's seat, there was no reason to rush things. I might as well relax and enjoy the ride. "I'd like some time to discuss it with each of them."

"Of course. You should expect a call. Two calls, to be precise."

When I arrived at the office the next day, there were messages waiting from senior salespeople at IBM and HP. I wondered which of them was the better dancer.

7

The Partner

AS WITH MANY of the world's religions, it takes a bit of extra faith to fully accept that the IBM corporate culture is completely benign. It instills in its employees a sense that they are tough, fair, and among the chosen. Yet individuals are occasionally expected to do questionable things for the greater good, like a CIA operative ordered to defend democracy by assassinating a head of state. IBM has a tendency to create internal myths that protect its sense of order. If it is outwitted by some fast-moving upstart competitor, it is probably because someone didn't play fair. If it doesn't have the best products, it may merely be trying to protect its customers from unproven technology.

Like the Marines, IBMers are taught that they must be ready to do whatever it takes. The IBM "do or die" culture has evolved a curious strategy for dealing with its own occasional failures, a technique that might be called "do or lie." Rather than accept a painful defeat, whole divisions sometimes continue their work as though nothing had happened — as though products had worked, as though customers had bought them, as though employees' efforts contributed to the bottom line. Company executives simply declare that the battle is still raging and they are still in it, like the last desperate days of the Vietnam War. OS/2, IBM's answer to Microsoft's Windows, is a prime example of the "do or lie" technique. Well over a billion dollars had gone into developing this operating system, an investment that continues today with no realistic possibility of a return. The technique works well as long as enough money pours in to buoy up moribund projects. But by mid-1989, IBM was so laden

with the walking dead that its highly profitable mainframe monopoly could no longer keep the rest of the company afloat. The problem was that nobody knew it yet.

Within hours of Norm Vincent's passing the word, both IBM
and HP had arranged to visit our modest offices to discuss potential
teaming arrangements.

IBM showed up in force: close to a dozen people crowded into
our conference room, led by Al Johnson. As the most senior IBM
salesperson assigned to working the State Farm account, he knew
that his most difficult task would be convincing his own firm that
State Farm was serious about working with GO. "Do you realize that
this matter got all the way up to the MC?" Johnson said to me before
we got started. He was referring to IBM's management committee,
a handful of top executives who met regularly to resolve the most
important problems facing the company. Given IBM's mammoth
size and far-flung offices, this four-man corporate equivalent of the
United Nations mainly mediated territorial conflicts and coordi-
nated initiatives so colossal that they crossed divisional boundaries.
Most employees could work at IBM for a lifetime and never so
much as shake hands with people that far up in the organization.
Johnson was clearly impressed.

It soon became apparent that the participants were from differ-
ent provinces within the company and had little desire to cooperate.
The group included researchers from IBM's prestigious T. J. Watson
Research Labs who had worked on handwriting recognition for
years; development engineers from the entry systems division, an
enormous product-development organization responsible for IBM's
PC products; and, of course, the sales team assigned to keeping State
Farm happy.

The meeting took all day. I later learned that this was standard
procedure at IBM: a simple review takes one day, and if there is
something substantive to discuss, it requires two. In contrast, GO's
meetings were usually scheduled in fifteen-minute increments.

During a break, Al Johnson took me aside. "Just between you
and me, what's it going to take to get you to bid this thing with IBM?"

I wanted to play it a little coy. "The issue for me is to avoid
having to raise our next round of financing from other investors."

"And how much would you be looking for?" Johnson asked.

"On the order of ten million dollars," I responded. "We could take five as a loan and five as a straight equity investment."

He didn't blink. "And if I get this approved, will you notify Norm Vincent of your decision?"

"I believe the GO board would look favorably on such an offer."

By the end of the day, Robert and I were exhausted by the endless pointed questions and tired of watching them fight among themselves. Johnson spent much of the time attempting to convince his people that State Farm was, in fact, intent on using GO's technology and not theirs — a conclusion many found hard to accept.

By contrast, the Hewlett Packard meeting was a breeze. The earnest and plain-spoken head of the HP's Bloomington sales office assembled a small collection of knowledgeable engineers and account reps. Their questions were relevant; their team obviously understood the customer's needs.

Robert and I caucused following the meetings. "So, who do you think we should work with?" he asked.

I sat there and stared at him. "Well, let's see. We have one potential partner who's skeptical of our technology, divided on whether they should work with us, and resentful of our success with State Farm. The other one is cooperative, has no competitive projects to speak of, is headquartered down the road from us, and is anxious to work with us." Now we were both staring at each other. We knew what we had to do. In the final analysis, what mattered was not how much we liked the partner, but how much weight they could throw behind our bid to establish our operating system as a standard. In this respect, there was no contest.

Robert broke the silence. "OK, OK, it's IBM," he sighed.

A common misconception is that business deals are cold, arm's-length financial transactions between faceless institutions. In truth, business relations are much like personal relations: good ones are built on trust, respect, and mutual benefit, and bad ones are based on deceit, exploitation, and betrayal. Curiously, those closest to business transactions, the lawyers, are often the most oblivious of the subtleties of the process. Their professional aversion to taking risks is at odds with the essence of business, which is placing shrewd bets that may return more than the wager. They also tend to confuse "the deal," the working understanding between two parties, with "the contract," the written words that attempt to capture that understanding at a point in time. Words are good for capturing some things, such as the rules of chess, but not for others, such as how to ride a bicycle. What makes deals work are not the written words but the personal relationships between the individuals charged with making them work. And that's why it's very difficult to do business with IBM.

Rather than empowering the responsible party to make a deal, IBM assigns a professional negotiator, who usually knows or cares little for the substance of the agreement but has absolute authority over all of its details. The negotiator begins by assembling a list of interested internal constituents, all of whom are free to add new requirements at any time or block some minor concession if it doesn't suit their needs. The result is like watching a photographer attempt to corral a group of hyperactive children into sitting still for a picture: if there is any delay in snapping the shot, some new disturbance inevitably erupts.

IBM's deal-making culture developed out of one of the greatest legal battles in history, the thirteen-year attempt by the U.S. Department of Justice to break IBM's monopoly on the market for mainframe computers. In the end, IBM settled the suit by consenting to some minor restrictions — such as not announcing in advance products that might chill competitors' sales — but at a cost perhaps as high as $100 million. A legal staff of hundreds was employed over the duration of the dispute. This ordeal left deep scars on the IBM psyche, resulting in a preoccupation with secrecy, an aversion to putting commitments in writing, and an elaborate system of legal reviews for even the most routine business transactions. At IBM, every press release, speech, and product disclosure is subject to legal approval, leading to interminable delays. This culture is unsuited to the fast-moving environment of the personal computer business.

After agreeing to join forces with IBM, I expected to develop a working relationship with Al Johnson, sort out the details, then draft and sign an agreement. But now the only person who would return my phone calls was a man named John Kalb. And to my surprise, Sue King, one of the more critical members of the original visiting team, was put in charge of the project.

Kalb, thin and slightly stooped, a heavy smoker, with a wave of brown hair and a soft voice, set about to create a Byzantine agreement between our two companies. He approached his work with the meticulous care and uncompromising zeal of an artist, seemingly with little concern for deadlines, cost, or consequences. Under his dedicated stewardship, an army of lawyers, accountants, security officers, escrow agents, and file clerks labored long and hard to create a masterpiece of legalese. Kalb engaged the top New York law firm of Cravath, Swain & Moore to augment his own in-house staff. He resisted my pleas to get the deal done and let me return to managing the company.

Kalb began by reinterpreting the basic structure of my proposal: the $5 million loan was to be secured by GO's intellectual property, which meant that if we failed to pay the loan back, IBM could take possession of our software, product designs, and patents. We were to receive the other $5 million as an investment upon the signing of a "comprehensive agreement" in which IBM might relabel and sell GO's products, provide access to proprietary advanced technology, and contract to do our manufacturing. But, of course, there had to be a thorough review of our technology and an approved internal-development plan before IBM would commit itself to the final agreement. "We can make you the Microsoft of pen computing, with IBM as your banker," Kalb would say when I voiced an objection.

The next step was a series of review meetings with Sue King and her hastily assembled team. In contrast to Kalb's grand visions, King saw us primarily as another supplier — one unaccustomed to meeting IBM's demanding requirements. She protested that our schedules were too aggressive, our project milestones were insufficiently detailed, our test plans were inadequate, our product designs weren't rugged enough. As hard as she was on us, though, her willingness to do battle with her superiors to promote the project helped get the attention and resources we jointly needed to succeed.

Morale at GO began to flag. A seemingly endless stream of calculator-toting IBM engineers would arrive for leisurely tutorials on object-oriented programming and pen interfaces. The visitors would take our engineers to lunch, where they would darkly confide that there was no way IBM would ever decide to back an operating system other than OS/2. As one of their minions warned me, "We can blacken the skies with planeloads of curious tire kickers."

The most comical of the visitors, perhaps because of their gumshoe demeanor, were from the IBM security forces. In anticipation of providing us with confidential documents that never actually arrived, they came to review our security procedures — which were nonexistent. They designated a special locked room for IBM confidential material, containing an even more special locked file cabinet. Only I was to have the single key, and anyone allowed access to its contents had to sign a register attesting to what they had read. They even required us to place a chicken-wire barrier above the tiles of the drop ceiling, in a preposterous attempt to impede any intruder trying to crawl through.

Despite her lack of faith in our ability to deliver, Sue King forwarded our tentative delivery schedule for preliminary hardware prototypes to John Kalb. At our next meeting, he changed the rules again. As soon as I saw the new language, I exploded. "John, what the hell is this new appendix about breaking the loan into four installments, to be paid as we deliver working units to Sue?"

In response to my exasperation, he acted as if this modification were as innocuous as changing a comma. "She is concerned that you may not be able to deliver on schedule. This is just a way to focus your attention on her needs."

"But speculative engineering schedules have no business as part of a contract! We may need to change things around on a moment's notice. For that matter, Sue may need to. Do you want us to have to call our lawyers and execute a contractual addendum each time?"

"You'll have to see Sue about that," Kalb said.

As the negotiations dragged on, I came to understand that GO was laboring under the cloud of what IBM saw as an unfortunate misstep: its own operating-systems agreements with Microsoft. IBM believed that Microsoft was a monster of IBM's own creation, an arrogant, undeserving upstart that had temporarily gained control of a key piece of IBM software. Although IBM couldn't seem to get the better of Microsoft until their current agreement expired, it could certainly avoid making the same mistake again with GO. If IBM was going to be instrumental to GO's success, it surely wanted a high degree of control.

Kalb introduced a series of contractual principles that sounded like articles from the Declaration of Independence, but in fact were little more than a polite way of saying that GO's existence would henceforth continue at the discretion of IBM. There was the principle of linkage, for example: any infraction of any contract with any party, no matter how minor — such as missing a lease payment on a copier — gave IBM the right, under the security agreement, to call its loans and seize our intellectual property. Worst of all was the principle of freedom of action: regardless of what we were compelled to do for IBM, it would not agree to any specific restrictions on its future behavior, such as engaging in competitive projects, abandoning support for our operating system, or courting our customers.

By late October of 1989, we were burning more than a million dollars a month, in part owing to the cost of supporting IBM. At that rate, the loans — after we finished collecting them — would only get us through February or March of 1990, which was barely long enough to raise money if the IBM deal fell through and I started immediately with other investors. It was now essential that I close the agreement without delay, whether I liked the terms or not.

Just when I thought we had everything nearly settled, we turned our attention to the details of the intellectual-property security agreement. I was no stranger to legal work, but nothing in my experience prepared me for what was to follow. This innocent-sounding idea, which I assumed would merely be a clause in the loan agreement, turned out to be the most complex free-standing legal document I had ever encountered. Hundreds of single-spaced pages in length, it described in microscopic detail a convoluted tango of onerous procedures. We were obliged to deposit to an escrow account a copy of every program, design document, drawing, sketch, specification, rendering, screen art, chip design, schematic, parts list, example program, tutorial, slide, programming note, erratum, guideline, help screen — whether handwritten, printed, or electronic — along with any copyrights or patents, until the loans were fully repaid. In addition to the materials for the escrow, the IBM lawyers insisted that we send them a list every quarter of the names and home addresses of all personnel who worked on the project, identifying their specific contributions, accompanied by a "certificate of originality" signed by each person, attesting that they actually did the work themselves. To amuse myself, I would occasionally invent imaginary types of intellectual property, which the lawyers would proceed to vehemently argue must be included in the agreement.

I reviewed the escrow requirements with Robert. "Jesus, you've got to be kidding," he said. "What the hell do you think we should do?"

I knew there was no way to hammer this back into something realistic, but I was also confident that it wasn't ever going to matter. "Once every so often you should send a backup disk and the documentation over to the escrow agent," I told him. "Who knows? Maybe someday we'll have a disaster and be happy we saved all this stuff." I couldn't have guessed that we were only days from such a disaster.

Despite my cavalier attitude, I was actually quite worried about the tar pit we had fallen into. I decided to call David Liddle, our new director, first thing the next Monday for his opinion. He returned my call the following afternoon from a pay phone in the basement of a hotel in Palo Alto, about ten miles south of our office. Ironically, he was just finishing up an all-day meeting with some people from IBM.

He listened carefully to my concerns, then gave a characteristically colorful response. "It's the sorcerer's apprentice."

"Huh? You mean like Mickey Mouse in *Fantasia?*"

"Yep. There's no one responsible in charge, so things just keep getting out of hand on their own."

"What should I do?"

"We need to get an adult to supervise the process," David said. "I'll call Jim Cannavino and see if he's willing to step in and straighten things out." Jim Cannavino was the general manager of a now defunct IBM division known as PS-LOB — personal systems line of business. David had become acquainted with Cannavino because David's company, Metaphor, also had a partnership with IBM.

"Thanks, David, I really appreciate your help, as always."

There was a long silence. Then he said, "Do you feel that?"

At first I wondered what he meant. It sounded as though he was questioning my sincerity.

"Feel what?" I said.

"We're having an earthquake."

"Nope, don't feel it up here." Then my chair started to vibrate ever so slightly, as though someone were rolling a loaded cart down the hallway. I instinctively looked at the plant on my windowsill, to see if the leaves were fluttering. Once or twice a year we had a perceptible earthquake, which was easily missed except for this subtle telltale sign. "Yep, now I can feel it."

"Wow, this is a really big earthquake," he said.

"No it's not," I replied.

Then the line went dead. I figured he would call back in a minute, but then realized what was happening. The quake's epicenter was to the south, and the waves were rushing northward. But David, in the hotel's basement, stood on firm ground, and I was on the fourteenth floor of an office tower built on landfill.

I grabbed the corners of my desktop to brace myself just as the first shock wave hit. It felt like a surreal explosion: no sound, no source, just a sudden jolt that pitched the entire building forward about a yard, as though it had been whacked on the other side with an enormous rubber mallet. This massive steel and concrete monument, topped with a postmodern pediment like a tombstone, leaned toward the bay at an impossible angle. From the floor-to-ceiling window in front of me, I could look down at the receding slope of the lower stories, which left me floating in air just outside where my office used to be. I was so startled by this sight that I didn't think to move.

Then the building lurched back the other way, jerking my chair from under me and rolling out the side drawers of my desk. I stood up, still clutching the desktop for support. There was no deep throaty rumble like you see in the movies, no hysterical screaming extras running for cover. There was only a horrible metallic groan that reverberated up and down the structure as the enormous steel skeleton danced to keep its balance. The building swung back and forth three or four more times, each time leaning a little farther and rolling the rubber base of my chair around like a bumper car until I managed to snare it with my left leg.

On the street several hundred feet below, I could see the ground swell up and down, bouncing the helpless cars around like corks in open ocean. The streetlamps, arranged in a graceful row of slender aluminum shafts with outstretched arms, whipped to and fro like a chorus line.

Then, as abruptly as it had started, the mayhem stopped. The scenery was frozen again, but there was one problem: I was still thrust precariously out toward the bay. For one crystalline moment, I hung suspended in space and time, wondering if by some seismic serendipity the entire building might actually fall over, ending my existence. As if to answer my question, the structure crept back to its neutral position with a deep, soulful gasp.

I let go of my desk and looked around. Miraculously, there was no obvious damage to my office, though the power was out. I ran out the door and called down the two adjacent hallways, "Is anybody hurt?" People were emerging from their cubicles and offices, dusting off the snowy white powder that had flaked from the ceiling tiles.

"Everyone's OK over here," Kevin called from his office at the adjacent corner of the floor. "All clear on this side," someone yelled from the other direction.

Robert emerged from his office, which was next to mine. "Wow" was all he said.

"The server — check the server!" Kevin yelled from down the hall.

"Oh shit, the server!" Robert echoed. This was the nerve center of our local area network, the main electronic repository of all our work. If it was damaged, we could be seriously set back. Not only would we lose everything since our last backup, but it could take weeks to reconstruct its directories and files. Kevin beat us to the small utility room where the server was housed. The folding table on which it sat was still standing, but covered with fallen manuals and other assorted debris. The screen that normally displays its status was blank, even though it had switched over to backup battery power. Kevin examined it with the professional touch of a physician. "It's OK," he announced. The assembled crowd gave high-fives all around.

We immediately vacated the building, in case of aftershocks. When I got in my car, I had my first indication of just how serious the damage might be. I turned on the radio and found that most stations had been knocked off the air. As I inched my way back to the city, through streets with no working stoplights, I suddenly felt ashamed of myself. I had been secretly thrilled by the solitary peril of that final, magic moment.

Within a few days, power was restored to the building and we returned to work. There was a message from John Kalb waiting on my voice mail, asking me to call him back right away. I immediately returned the call.

"Cannavino wants to see you," he said. His tone was grave and deliberate, as though delivering the word of God. Clearly, David Liddle had gotten through. I was pleased, not only because he had pierced the IBM armor, but also because I hadn't heard from him since our aborted phone call, and it meant he was uninjured.

"Cool. When can I see him?"

"There will be two meetings. The first of these will be for one hour, in the IBM executive trailers next to the main conference hall, the Tuesday of COMDEX, and the second one will be two days later, at IBM worldwide headquarters in Armonk, New York." His formal phrasing made COMDEX sound like a week-long religious holiday. Perhaps this was his way of expressing disdain for my casual response to this rare opportunity, or maybe he was just annoyed at being cut out of the loop.

"What does he want to talk about?" I asked. "Should I do any special preparation?"

Kalb paused as though deciding whether it was safe to discuss this over the telephone. His voice took on a mysterious tone. "Come alone. Make no special preparation. Just be ready to talk," he said. "And there's one other thing."

"What's that?"

"Don't tell Sue King. In fact, don't tell anyone at IBM that this is happening."

I caught an early flight to Las Vegas that Tuesday. Much of the road into town was nothing but barren, empty desert, dotted with billboards tempting pious visitors with forbidden delights. Even by midmorning, a relentless dry heat caused these lures to shiver like devilish mirages.

I arrived at the convention center thirty minutes early, only to consume the extra time walking from one end of the building to the other. To my amusement, I passed through the same door that John Doerr and I had used to escape the previous year, and followed the signs to IBM Conference Trailer B. I was ushered into a small white room with orange-beige shag carpeting. A folding conference table and six stackable chairs filled the room. In one of the two small windows, a large air conditioner slaved loudly in a futile attempt to stave off the desert heat, making conversation difficult.

Jim Cannavino entered and greeted me. He was a compact, broad-shouldered man with neat, jet-black hair who looked and talked like an earnest plumber. We proceeded to use up about forty-five minutes of our hour getting to know each other. I talked about my vision for GO and pen computing; he told disarming anecdotes about himself, his career, and the company. "I started by repairing keypunches," he said, "and got to where I am by bitching about what a lousy job my boss was doing, until they gave me his position. I bitched about the PC division, and here I am, heading it."

When he noticed that my time was running out, he got to the point. "I want you to know that I really like your stuff. Now IBM" — he gestured expansively, like a king surveying his land of plenty — "does not have a good history with small companies as partners. We tend to bog them down in bureaucracy. But the partnership with GO is one that I want to see succeed. I've had a look at the plan our people are putting together, and I'd like to know what you think of it."

He stopped and looked at me. I assumed that he was stating this rhetorically, and that after a dramatic pause, he would continue. I couldn't imagine that he actually wanted my opinion of this plan, when his own staff had at best taken no steps to share it with me, and at worst were actively trying to keep it from me. Since they needed a great deal of our help to make the project credible, of course, I had a pretty good idea of what they had in mind, and thought little of it. Unconvinced of the idea that pen computers were really a new class of machine, they were planning to add our technology to OS/2 and shoehorn it into modified laptops, which would have a keyboard that could be detached when you wanted to use the optional pen. But this was fine with me, because if they followed through, it would leave the field wide open for GO's lightweight tablets, which customers would vastly prefer.

"Its OK, you can level with me. It'll stay between us," he said, encouraging me to speak. Then it dawned on me that this was why he had insisted on secrecy, so that his people wouldn't be able to lobby me into supporting their plan. Maybe Cannavino wasn't such a bad guy.

"Frankly, Jim, I think your team doesn't quite get the point of pen computers yet," I said. "They want to build hybrid laptops with pens, and perhaps that will work. But personally, I think they may not be taking the best path."

He leaned toward me and lowered his voice. "So you believe they're wrong, then?"

"I suppose so," I said hesitantly. I wasn't sure how he was going to react if I disagreed with the recommendation of his staff.

He sat back and exhaled loudly. "Good. That's why I asked you here, to get my organization off your back. All they're going to do is throw rocks in your backpack, weigh you down until you sink. But I don't just want your technology, I want you to call the shots."

I felt like a man stranded in the desert who was promised an ice-cold drink. I was so drained after enduring months of IBM negotiations that I could almost taste the freedom of pursuing the project unfettered by their demands. He could see this reaction in my face and immediately went for the close. "I'm ready to take a risk on you. If I have my way, I'd buy you right now, and then let you do what you want. How is that?"

I was totally unprepared for this, and really needed some time to think. Working for IBM wasn't the cruise that I had signed up for, but that probably wasn't the point. The question was what's best for the employees, the investors, and ultimately the customers. I was speechless.

"Why don't you think about it and we'll talk again on Thursday in Armonk," he said.

"That sounds good to me. I'll see you then."

As I walked from the trailer back to the conference center, I was uncertain about the right course of action. I thought about the team back at the shop, wondering whether they would think of an IBM acquisition as selling them out or setting us on the path to success. It was a very tempting offer, but it appeared that the price of freedom, financing, and IBM's support was to sell my soul to Jim Cannavino.

My next move was to talk to John Doerr. I found him in a corridor of the Las Vegas Hilton, and we looked around for a place to sit. There weren't any chairs, so John dropped to the floor and crossed his legs. I hunkered down and replayed my conversation with Cannavino.

John looked exasperated. "Jerry, do you want to sell out?" he asked.

"No, not particularly. I don't think I have the patience to work for IBM."

"Right, then don't give in to Cannavino. That's not what we're in this for." John's voice grew louder. "We've got work to do, and we can do it without them if we have to." He was working himself into a frenzy.

A hotel guard patrolling the hall stopped and eyed us suspiciously. There we were, two guys in suits having an animated discussion while sitting on the floor. He leaned over and said, "Are you the ones waiting for the girls?"

John didn't seem to understand what he meant.

"Nope, sorry," I said.

As the guard walked on, John said, "Just tell Cannavino it's not for sale."

Sometimes I wasn't sure what motivated John. It certainly wasn't money. To most investors, the prospect of a quick sellout to IBM would be irresistible. But for John, investing in high technology was a calling, and monetary reward was the byproduct of a successful mission. The higher goal was to create something enduring, a growing enterprise that delivers products, employs people, and enhances the wealth of its stockholders.

Since I would be arriving late in the evening at New York's Kennedy Airport after a full day of meetings at COMDEX, I arranged in advance for a car and driver to take me to the Holiday Inn in White Plains.

I scanned the row of limo drivers waiting in the baggage claim area. They were all husky, tough-looking men with greasy hair, standing shoulder to shoulder and holding up little signs bearing the names of their charges, like an accidental police lineup. I pointed to a fellow with leathery skin and a frayed polyester tie. "You're the one."

"Follow me," the driver said. "I have a message for you in the car." When we got under way, he handed me a piece of paper. "Call home," it read.

This was extremely odd. There wasn't anyone at home, so I deduced that it must have meant to call my parents. After all, only they would think that a child, grown and gone for twenty years, who had never lived in their current house, would call their place home. How they had tracked me down I couldn't imagine.

"You can use the car phone," the driver said. I dialed Palm Beach and a sleepy male voice answered.

"Hello, Father?" I said.

"Jerrold!" Only members of my family called me by my full middle name. "Where are you?"

"I'm in a limo in New York on my way to White Plains."

"A limo!" he said. "Fancy. What are you doing there?"

"I have a meeting at IBM headquarters in Armonk tomorrow."

"What about?"

"Well, they want to buy the company."

"That's great."

"Yeah, but I don't want to do it."

"Don't be stupid," he bellowed. "Sell it while you can!"

"Thanks, I'll think about it." I didn't want to get into a long discussion, and besides, he wouldn't understand.

"Grant me this one last wish," he pleaded.

"Dad, I said I'll think about it. I'm not the only one involved, you know." Then I realized what he had said. It didn't sound like his usual histrionics. "Is everything OK?"

"Fine," he said. Of course, this made no sense. If everything was fine, why was he tracking me down in New York in the middle of the night? So I waited.

"Listen, I've got some bad news," he said. "I'm going to die."

"You mean for real?"

"Believe me, I'd fake it if I could."

"Come on. You're kidding, right?" This wasn't something a dying man would say, even my father. But then again, humor is the way many Jews deal with tragedy, especially death.

"No, really. I've got cancer of the liver. It's advanced."

I couldn't believe it. For the past three years, I had been totally wrapped up in my work, so much so that the rest of my life — family, friends, recreation, reading, music, anything other than GO — had virtually ceased to exist. I had always justified this by believing that my life was merely on hold, slowed down like the body of a hibernating bear, to be revived on cue with the coming spring. I had ignored birthdays, anniversaries, holidays, and the other events by which people mark the passage of time, figuring that I would apologetically hop back into the parade later. But now I was faced with the stark reality that life went on without me, ready or not, and once these landmarks were passed, there was no turning back to revisit them. Wrapped in a cocoon of meetings, memos, and milestones, I had let my family drift away like a neglected water lily in a gentle pond, only to find it withered and out of reach when I needed to retrieve it. It's one thing to skip a birthday present; it's another altogether to skip your father's life.

I didn't know what to say. "Christ, that's terrible. I'm so sorry."

"Not as sorry as I am."

"How long have you got?"

His voice grew weak. "A few months maybe. It doesn't hurt or anything. I wanted you to know, so you can take care of your mother."

"How's she taking it?"

"Pretty well. But I think she's upset about having to find another golf partner." He wasn't a particularly introspective man, so I expect that the metaphorical interpretation was not intended. She was seventeen years his junior, only sixty-five years old, and her mother — my grandmother — was still going strong at eighty-nine.

"Look, I'll turn around and come down there right away." I knew that missing the meeting with Cannavino could be disastrous. GO would soon be out of money again, and there was no telling if I could get things back on track. Startups are like babies — you can't just take a break when something important comes up. But I knew the right course of action was to go to his side, even if it meant losing the company.

"No, that's silly, go to your meeting. What are you going to do, sit around Palm Beach for months waiting for an old man to kick the bucket? Remember, work comes first. Come when you get a chance."

He had a point, so I reluctantly agreed. "OK, I'll come and visit as soon as I can."

"Close the deal first," he insisted.

"Dad, I love you very much." This was something we rarely said, though we both knew it to be true. I was determined to change that in my own family, when I had one.

"I've got no regrets," he said. "I have a family that I'm proud of, and there's really nothing else I want to do. Besides, what's left to live for? My golf game is terrible." He hung up.

I stared at the floor, listening to the tires thumping over the seams in the concrete highway like an irregular heartbeat, wondering about the choices I had made in my life. But I had to put this out of my mind, at least until after the Cannavino meeting.

The next day, I took the long ride to Armonk. The weather was gray and overcast; sharp gusts of wind badgered the trees into giving up the last of their leaves. The IBM facility was in the middle of nowhere, forty minutes from the nearest hotel. When I arrived at the sprawling campus, it was drizzling. Identical buildings seemed to stretch for miles across the undulating grassy hills, distinguished only by discreet road signs with a confusing jumble of letters, numbers, and arrows.

The visitors' entrance in the main building was sixties industrial chic — clean, cold, and antiseptic. Wide expanses of glass and steel formed a cavernous chapel with a raised central pulpit where the receptionist sat. The only other furniture was two Bauhaus-style benches of chrome and leather. A lone Japanese visitor occupied one of them, in the slumped posture of an underling; I sat on the other to wait for John Kalb, who was to chaperone me for the day. I noticed that the bench's upholstery was frayed and worn. A pail in one corner of the room patiently collected rainwater dripping from a leak in the roof. The place looked deserted.

A short time later, Kalb arrived to sign me in.

"Is this corporate headquarters?" I asked incredulously.

"This is *worldwide* headquarters," he said, grinning. "This is where the rubber meets the sky."

We walked down an endless series of corridors, most of them lacking any signs of life. At one particularly desolate intersection, a television played to an nonexistent audience. There wasn't a chair in sight, even if someone had decided to stop and watch for some reason. Yet this electronic sentinel droned on. "IBM internal news network," Kalb explained. "I believe it's worldwide, operating around the clock."

Kalb led me down an escalator to an executive dining room decorated with dark wood paneling and woven beige wallpaper. A picture window looked out on a pastoral scene of bare trees on the banks of a pond. A single duck paddled around aimlessly, confused by the changing season. On one wall of the room a small sign hung below a picture hook: "Art is temporarily removed." There was a call button for the waitress, but it didn't work.

The meeting was scheduled from twelve-thirty to two-thirty. We waited over an hour for Cannavino to show up, then Kalb called his last known location. Cannavino was still waiting to present his yearly plan to a review board; no telling when he could break free. Just when we were ready to give up and leave, he arrived, distracted and hungry.

I made my pitch as he wolfed down his lunch. "Jim, I've thought about the possibility of IBM buying us out, and I don't think it's the way to go. For one thing, to build a market in pen computers, you need a credible independent party to license the operating system to other manufacturers. For another, I'd rather see your money go into building great products than into lining the pockets of GO stockholders, at least at this point."

He was visibly disappointed.

"I'd like to propose something different," I said. "Take a twenty-five percent equity position with an option to buy more later. License our software technology on 'most favored customer' terms, and I'll give you a partial credit against IBM royalties for all licensing revenues I get from others. That way, you can participate in the growth of the market beyond only your share. Then we can jointly develop the hardware, which you can manufacture and only GO and IBM can sell."

"You know, I could do this alone if I wanted to," he said.

"True, but I can offer you quick time-to-market, and you know how important that is." Shortening product-development cycles was all the rage inside IBM, and I was trying to push one of his hot buttons. He wasn't looking any happier, so I tried another button. I knew that he was in the middle of a difficult face-to-face negotiation with Bill Gates over their cross-licensing agreements. "If you back me, I can get you into this market way ahead of Microsoft." I wasn't at all sure what, if anything, Microsoft was thinking about in this area, but it didn't really matter, because Cannavino took the bait.

His eyes narrowed. "What does Gates think of you?"

"He's scared enough to want to beat us to the punch, and without your help, he just might." I was bluffing, though this proved to be a prophetic fabrication. But our time was almost up, and I knew I had only a few minutes to light a fire under Cannavino if GO was to survive. This put him over the top.

He turned to John Kalb and said, "OK, work something out. I want this deal done in two weeks." Kalb, who had sat silently throughout, was surprised at this sudden call to action. He knew that two weeks was grossly unrealistic, particularly now that we had discussed a completely new structure for the relationship.

"Jerry, thanks for coming," Cannavino said. He threw on his raincoat and rushed out to his next appointment.

Kalb walked me back to the main entrance, saying very little. I offered to get the ball rolling by writing up a summary of the proposed relationship and faxing it to him. "That would be a help," he said, scratching his head. I left for the airport.

The next morning, I got up early to work on the proposal, then faxed it to Kalb before nine, knowing this would give him half a day to work on it in New York. I needed to get things going fast, not only so that GO wouldn't run out of money, but because I was frantic to get down to Florida to see my father.

Kalb called me at home two days later, on a Sunday afternoon, to talk about the deal. But he'd had little time to sort out his thoughts and seemed reluctant to react positively to anything. "I don't like your proposal, but I'll think about it and call you back on Monday," he said.

He didn't. Finally, I caught up with him late on Tuesday, and he explained that he'd been tied up with other matters. This seemed strange, given the urgency that Cannavino had expressed at our Armonk meeting. Kalb asked for a few more days to study the situation and promised to get back to me before Thanksgiving, which was the following weekend.

Then, to my utter amazement, he dropped off the face of the earth. Cannavino's deadline came and went without note. To this day, I haven't got a clue as to what happened.

But desperate as I was to close a deal, I had something more important to do.

Once I realized we weren't going to get the IBM deal settled quickly, I called my staff together and told them of the problem with my father. "I don't know how long I'll be gone. You're not going to want to call me, but I want you to know that it's really OK to phone me any time of the day or night." I knew that business would grind to a halt if I was unavailable to approve job offers, sign certain papers, and keep working on IBM. Only Robert and Kevin really understood how critical the situation was — every day we didn't close the IBM deal, we risked running out of money. But they respected my decision and did what they could to fill in. I took off for the airport.

The common practice in Palm Beach was for doctors to release terminally ill patients to die in their own homes, if at all practical, rather than in the cold and unfamiliar environment of the hospital. This placed a significant burden on their families, however, who were usually poorly equipped, either physically or emotionally, to deal with the bodily requirements of the dying.

As soon as I got to the house, I went straight to the master bedroom to see my father. Mother had converted the room into a kind of command center. At its heart was an adjustable hospital bed, surrounded by tables covered with monitors, catheters, and other rented equipment, as well as an extensive collection of medicines and painkillers. She had hired a nurse to relieve her during the day, and when I first arrived she was handling night duty herself — a foolish economy that we quickly corrected.

My father was deteriorating rapidly, unable to stand up or relieve himself on his own. You could actually see his entire skeleton through the skin, which had turned thin and waxy, and was mottled with the blue and brown remains of blood vessels. He was clothed in an adult diaper and a T-shirt. A catheter led from under the diaper liner to a clear plastic bag hanging from a railing around the bed. Every so often, small sections of gooey yellow fluid moved a few inches toward the bag, then came to rest again. Whoever was on duty periodically swabbed his lips with a wet tissue to keep them from cracking while he slept, and occasionally fed him a small quantity of water from a dropper when he got thirsty.

I didn't know a human being could become so frail, and was shocked to find him in this condition. The last time I had seen him, he was trying to persuade my mother to play nine holes with him, because none of his regular partners were foolish enough to go golfing in the rain. As my grandmother once observed, "It never rains or snows on Murray's golf course."

My brother and two sisters were already in town, and had been for some time. At that moment, my little sister Amy and her husband, Hal, were at the house, preparing to go out to dinner with some friends as a well-deserved break. They were standing at the foot of the bed. My mother was on one side, I was on the other.

"Murray, Jerrold is here," she whispered in his ear.

I took my father's hand. "Hello, Dad," I said. He slowly turned his head and smiled. His eyes darted around, but they didn't seem to focus. His fragile fingers curled around mine. Then his other arm rose slightly, appearing to point at Hal. He parted his lips and inhaled slowly, in preparation to speak. My mother and I leaned in.

"Now I can die," he rasped weakly. I was badly shaken.

Oddly, my mother was upbeat. "Hal, he says he likes your tie."

I wasn't going to correct her. That was the last thing I ever heard him say, for he soon fell into a coma.

During the next few days, I learned a new respect for my mother. I had mainly seen her as a rather self-indulgent socialite, always complaining about what a strain it was to have to go out so many nights a week, or about how many people to whom she owed return invitations. But for the weeks leading up to my father's death, she showed a degree of selfless devotion to his needs, no matter how distressing, that I had never witnessed before in my life.

One morning, I answered the phone. "This is Doctor Mash, of the Miami Brain Bank," said a pleasant woman's voice. An implausible name if I ever heard one, but she sounded sincere enough. "I know this is a difficult time, but I have an important request to make. I understand that your father is in the early stages of Parkinson's disease." This was true; it was the main reason he had been so upset about his golf game. "We do research on this subject and would very much like to ask if we could remove his brain for study after he dies."

I was incredulous. I remembered my father's long-standing advice on the virtues of a good education: "They can take away your money, they can take away your business, they can take away your house, *but they can't take away your brain.*"

"This is a joke, right?"

"No, I'm dead serious," she said. A poor choice of words, I thought. "We will arrange to pick his body up, transport it to our facility in Miami, and have it back to the funeral home in Palm Beach within a few hours."

I decided to risk asking my mother. To my surprise, she agreed. I made the arrangements, and she signed the necessary papers.

His condition continued to worsen. Inexplicably, we had a hard time getting a doctor to come to the house and check on him. We finally persuaded the son of one of his bridge buddies, an M.D., to take a look. He came by early in the morning, on his way to work. "I should just warn you that he could remain this way for several weeks. Don't assume that this will be over too soon," the doctor said after his examination. He left, and again we were on our own.

By this time my father was unconscious. My mother took me aside. "Jerrold, we have plenty of help here now. Why don't you go and take care of your business? I'll call you back as soon as it's over." Once again she surprised me. Having overheard many of my phone calls to the office, to the directors, and to various IBM offices, she was beginning to understand that something was very wrong.

"I couldn't do that," I replied.

"Don't be silly. After all, it's what your father would have wanted."
I went in to see him again. He was completely unresponsive.

"You'll call me right away if anything happens?" I asked my mother.

"Absolutely."

I booked a flight for late that afternoon. When it was time to leave for the airport, I called ahead to check the departure status. To my amazement, the plane had already left. "Sorry sir," said the agent. "There was one at four-oh-five, but there's nothing at four-fifty." For the first time ever, I had missed a flight.

It turned out to be one of the luckiest things that ever happened to me. Within the hour, my mother called me in to look at my father. She was concerned because his breathing had changed. It was now a labored, steady pattern of deep breaths, as though he were gasping for air. Every few minutes he would abruptly stop, then convulse as if slapped on the back, and begin again. It seemed strangely familiar. Then I remembered that this was the same thing that happened to my cat, and for the same reason: kidney failure. My mother had commented earlier in the day that the urine bag was filling much more slowly than usual. I suddenly felt comforted, more in control by virtue of my other experience with death. I knew what to expect.

"This is it, Mother," I said. "Call everyone together." Soon my family gathered at the house. We sat quietly in the library, taking turns monitoring my father. The periods when his breathing would cease grew longer. Each time, those present would wonder whether he would start again. Eventually he stopped altogether, but there seemed to be no way to know for sure. My two brothers-in-law and I were the only ones in the room at that moment.

"Do you think he's dead?" Hal asked. None of us had ever been posed this question before — to determine that intangible, illusory boundary between life and death. The gruesome possibility of declaring it crossed too soon would be more than we could bear. This dreadful conflict hung in the air like a damp chill. Then I realized they were waiting for me — as Murray's son, I was the only one with the moral authority to declare an end to the nightmare. I looked at my father's motionless body one more time, then nodded my head once, as though ending a prayer. They both nodded in response, making it official: he was gone.

I called the removal service, as I had arranged with Dr. Mash, while the others returned to the library to deliver the news. Again, my mother surprised me. She didn't cry or break down. Instead, she seemed relieved, as were the rest of us. The strain ended at last, the uncertainty over. But she had one last request, to go in and say goodbye. I accompanied her.

She entered the room, as if visiting a sacred shrine, and straightened his sheet, saying nothing. I could see only love in her eyes, no hint of grief or any other emotion. She was completely at peace. At that moment, I understood what marriage really meant — and realized that I might now be ready for it.

Then she was distracted by his hand, which was extended over his head, gnarled and pointing. She held it lightly. "I always loved his hands," she said. After a few seconds, she looked up. "I'm done," she said abruptly. We returned to the library.

While we sat waiting for his body to be picked up, the phone rang and I answered. Everyone looked at me to see who it was — a ritual that occurred every few minutes during the day. It was GO's acting VP of marketing, calling for approval to hire a new director of corporate communications we had been recruiting.

"Ah, maybe this isn't the best time to discuss it," I said in a low voice. "My father just died. How about if I call you back in the morning?" I hung up the phone.

The scene seemed so ridiculous that mother started to giggle softly. Then my sisters joined in. "Could have been worse," whispered my sister Amy. "At least it wasn't for him."

"Why are we so quiet? It's not like we're going to wake him up," my other sister added.

Now we were all laughing out loud. Then my mother spoke. "You know, it's kind of a shame that you die when you're old. This would be a lot easier to handle when you're young," she said. With this, everyone was in hysterics — just as the doorbell rang.

Standing on the front step were two men wearing the colorless, drawn expressions of undertakers. One was tall and stooped with a long, thin face; the other was short and bald. Both were as pale as wax figures, as though they came out only at night. "We're here for the deceased," the taller one said. Everyone was watching, trying to regain composure. But from our flushed faces it must have been obvious that we had been laughing, even if the men hadn't heard us from outside. They looked at us as though we were crazy.

The custom among Jews is to observe a period of mourning for seven days after the funeral. During this time, friends come to the house to pay their respects. Although my father was in no way religious, it seemed appropriate to honor this custom, which was expected by the Palm Beach community. This was the first time I got a chance to meet and talk with his circle of friends. What emerged was a picture of a man I had never known, one obsessed with breaking free of his ethnic roots to find acceptance in the broader society. He avoided synagogues, shunned Jewish fundraisers, and never spoke a word of Yiddish in the house, although I learned that this was his mother's native tongue. Despite his efforts, he was always confined by his background, restricted to joining the Jewish clubs, the Jewish businesses.

. I came to understand what he had tried to do for me — why we had moved from the Jewish suburbs of Westchester to the polyglot world of Manhattan; why he had sent me to something called Ethical Culture school on weekends rather than to Hebrew school; why he rarely spoke about our family history. He had molded me not in his own image, but in the image of what he wanted to be. He understood that much of his personal pain was caused by his own awareness of being different, and he believed that the less ethnic baggage I carried around in my heart, the better I would be able to break through these barriers. To achieve his dream for me, he had to let me go.

◫

"Jim Cannavino has asked me to step in and advise him on this situation."

I was relieved to find someone who was at last willing to pay attention to making a deal. The endless technical requests had continued unabated, but Sue King steadfastly refused to handle contractual matters — unless she wasn't getting what she wanted. It was now mid-January 1990.

The caller was Mike Quinlan, a long-time IBM marketing executive who had recently returned from a tour of duty overseas. He was apparently between assignments, and Cannavino turned to him to pick up where Kalb had left off. There was only one problem: Quinlan had little idea of what had transpired to date, and knew even less about GO. He had simply been instructed to work something out.

To hear Quinlan speak, you would think that IBM was an official arm of the Department of Defense. He used the military lingo of a Pentagon general, asking for background briefings, alluding to back-channel communications, and taking orders from the commander in chief. He had the broad smile and tanned appearance of a politician who got elected on the strength of his image and war record. I could envision him charging up a contested hill in North Korea, proudly carrying the Stars and Stripes aloft at his own peril.

Sue King assured me that I was very lucky to have him assigned, because he was a "*corporate* vice president." Like a bank, it seemed that everyone at IBM was a vice president, but she explained that this special title was bestowed on only sixty-four meritorious individuals, who were thereby afforded wide latitude to marshal company resources as they pleased.

Quinlan was quick to tell me that he was empowered to make a decision on the GO partnership and that he was a close friend of Cannavino, whom he referred to as Jimmy. He made it clear that all communications with Cannavino were to go through him: he would decide when and how Jimmy was to be consulted, if necessary.

Quinlan liked to conduct business in person, so I flew to the IBM entry systems division's headquarters in Boca Raton. This conveniently allowed me to stay with my mother, as Palm Beach was only thirty minutes away. The night before our first meeting, an IBM courier arrived to deliver Quinlan's opening proposal. It was frightening.

IBM was to get 50 percent of all revenues we received from parties other than IBM and licenses to all of our patents and copyrights. Equally outrageous, IBM's royalties to GO would be capped, meaning that after a period of time, IBM would have free use of our products. The promised $5 million investment was replaced with a $2 million prepayment of royalties. Gone was any hint of killing IBM's own project and backing ours, as Cannavino had suggested.

No sensible business person would consider these terms, because they were virtually guaranteed to put GO out of business. No one can afford to pay a sales tax of 50 percent and survive. I called Vinod

Khosla, one of our outside directors, who had a well-deserved reputation as a shrewd negotiator, and read him the IBM proposal. I held
the phone for a long moment while he considered what to do.

At last he said, "Come home."

"What?"

"They're not serious. There's no point in continuing the discussions."

"But that's suicide!"

"We can cut back to a small core of staff, which will stretch the
cash for a few more months while we figure out some other strategy.
Maybe we'll go into some other business . . ."

I figured I'd better call John Doerr.

"Sounds pretty bad, pretty bad indeed," John said. "You're going
to have to do something to turn them around. One way or another,
you have to make this deal. GO's survival depends on it."

I also sought out David Liddle. "The problem is that the cup-
board is bare," he said. "They aren't going to tell you this, but
someone discovered that the cookie jar is empty. IBM's last quarter
was a disaster — they're going to post a major loss for the first time
in decades and write off over two billion dollars in bad investments.
The word inside is they might even have layoffs for the first time in
history. If you want to make a deal, do it fast and make it cheap.
We'll have to find operating cash elsewhere."

I was stunned. The great IBM, whose international influence
rivaled that of a world power, was headed for a fall? I chuckled at
the thought of Silicon Valley being flooded with the résumés of IBM
lifers with obsolete mainframe skills, as though the British royal
family now had to find jobs waiting tables. But if IBM wasn't going
to be a sugar daddy for my company, despite all the big talk, I had
to close some sort of deal fast and seek other investors if GO was
to survive.

The following day, I was led to a conference room with Mike
Quinlan and a dozen supporting staff. We spent the first half hour
on introductions, explaining who was who and why we were all
there. It was clear that Quinlan was the only person present who
was free to express an independent opinion.

I began with a briefing on the nature of our business and the
market. This was the first time he'd gotten the complete story, as far

as I could determine. Then I started in on the terms, attempting to sweep them aside in a conciliatory manner. "Mike, let's talk about the problem you're trying to solve with this proposal. I'm sure you're not looking to put us out of business. What's the real issue here?"

"We want to accomplish three things," he said. "First, if we're going to make you a success, we want a piece of the action. Second, we have to be sure we can meet our commitments to customers if you disappear. And third, if this thing is big, we don't want to be paying you off forever."

I noticed that his points corresponded to the three main terms of his proposal: the sales tax, the intellectual-property rights, and the royalty cap. It soon became apparent that Quinlan wasn't wedded to the actual proposal, so I surmised that he had just given the same directive to an assistant and instructed him to draw something up. Since he'd made no personal investment in the terms, I decided to try to win him over with something that sounded even more dramatic.

"Mike, I think I can meet your goals in a much simpler way. I'm willing to give you an option, which you can exercise at any time, *to take our technology and do whatever you want with it.* If you exercise this option, you'll have to pay royalties on a declining schedule for five years, then you can do whatever you want without paying us a cent."

Quinlan's eyes widened. "Why would you want to do this?" he asked.

"I want to build a long-term relationship, and I'm willing to bet that we can continue to earn your respect and support every day. If this will help build the trust that's necessary for IBM to get behind us, then so be it."

Now his eyes narrowed. "And what do you want in return?"

"I ask only three things. First, I want to convert the five-million-dollar loan we already have into advances on royalties, in addition to the two million you are proposing. Second, at our option GO can walk away from the relationship as well. You'll get the source code, the technology you need to meet your commitments, but no rights to future versions. And third, you have to agree to continue to support our APIs."

From my perspective, a deal like this would have a number of advantages. It would get us out from under the intellectual-property security agreement. I wasn't worried about their taking the source code and competing with us, because I doubted their engineers' ability to make substantial improvements, much less to keep up with us. And most important, as long as we controlled the APIs, it didn't matter what IBM did with the code, since its efforts would only reinforce our own in recruiting applications developers.

Quinlan asked me to leave the room for a few minutes. I was hoping that no one in his entourage understood that the APIs, not the code itself, were the crown jewels.

"I'll take it," he said when I returned. "Except for one thing. I'll only convert three of the five million dollars in loans into advances. Two million stands." He wasn't going to be satisfied unless I agreed to his terms. I desperately wanted to kill the security interest, but I decided it was better to fight that battle another day.

"OK, let's do it," I responded. He smiled approvingly.

I spent the next two months in a mad scramble to sign the deal and collect the $2 million before our cash ran out. Whenever it came time to discuss details, Quinlan himself was often absent. He'd fly out to our offices, give a pep talk about the value of having Jim Cannavino stand up at the press announcement, and then depart to deal with other matters. He'd leave behind his negotiating entourage, who'd wear me down on the licensing terms, which were as onerous as those of the loan. By March, with only a few weeks of cash to spare, we signed the deal.

As I dropped Quinlan off at the airport to return to Armonk, we talked about the "joint walk-away" provision — our mutual right to terminate the agreement and go our own ways. But he didn't want to leave on this sour note. Standing on the curb at the departure level, his carry-on bag slung over his shoulder like a soldier's rifle, he smiled and shook my hand as though he were presenting me with a medal. "Anyway, you and I both know that it will never come to that. Can you imagine the damage to IBM's reputation if it tried to destroy a small, innovative company like GO? We simply wouldn't do that."

8

The Announcement

IT HAS BEEN said that a man lives two lives, one his own and one on the lips of others. The image makers of Hollywood have refined this concept by eliminating the original, leaving only the glittering reflection. But the engineers of Silicon Valley have once again built a better mousetrap: they have found a way to take appearance and transform it into reality. The trick to establishing a new computer product is not to make it great, but to make everyone think it is going to be great before it's actually delivered. Here's why.

All computer products are hopelessly interdependent, stacked like faces carved in an endless technological totem pole: applications designers exploit operating systems tailored to computers built around chips that implement protocols based on specifications developed in committees staffed by applications designers. Everyone in this evolving loop must guess which new technologies are most likely to provide a solid and enduring platform on which to build their own piece. The problem is that these commitments must be made long before the surrounding pieces are ever delivered, because the computer business moves so fast that waiting is suicide. So resources are allocated on the basis of promises and reputations, not observable facts — like oceanfront condos sold on the strength of a developer's rendering. But once a foundation is chosen, it is always easier to work around flaws than to start anew somewhere else. So the successful players in this elaborate guessing game are those who are able to persuade the most people to bet on them, because this justifies enough continuing investment to shore up their goods before their supporters defect to another platform. Insiders know the sorry truth: PR is as important as R&D.

In the computer business, getting your message heard is mainly a tug of war with a pack of skeptical yet gullible news hounds who make their living by tracking down stories before others do. Their job would be a simple matter of snooping and writing as quickly as possible, except for one nasty problem: different types of publications have different lead times — the interval between when a story is submitted and when it actually gets published. Some monthly magazines may require up to four months; a newspaper needs only a few hours.

This is the same problem that afflicts jungle animals, both predators and prey, who must find a way to share a common resource, such as a watering hole, without wiping each other out. As in the wild, a set of explicit but unwritten rules has evolved that makes it possible for everyone to survive: you don't attack a gazelle while it's drinking; you don't break a story on which another publication has arranged an exclusive.

None of this would matter to the people who make new-product announcements, except that no publication will print your story unless it thinks your product is news. As a result, you have to engage in an orchestrated plan of background briefings, lunches, press mailings, and conference calls, culminating in a newsworthy "event," so that everything can hit the street at the same time. You do this by notifying the press that an announcement of interest is forthcoming, scheduled for release on a particular date. By convention, all publications agree to honor such dates as long as they can get the story directly from you in advance. This is called a news embargo, and it affords you the breathing room to explain your product to all interested parties without blowing its news value.

It is the job of the PR firm of a high-tech company not only to know and obey these rules, but also to keep track of which reporters are hungry enough to break its client's embargoes. Since reporters seem to change jobs as frequently as migrant farm workers, maintaining this blacklist is a thankless and time-consuming task. The PR firm is your guide through this jungle, knowing who will feed you and who will eat you alive. What makes life hard for PR people is that in a pack, human reporters may not always be as trustworthy as animals.

In the early spring of 1990, GO's product development was heading for trouble. We had originally planned to release our first product by the end of March, but it had been repeatedly delayed, until the fall sounded more realistic.

Part of the delay was due to IBM. Whenever its engineers met with State Farm, they would impose a new set of requirements on us. If I objected, Sue King would sometimes threaten to exercise IBM's option to take the source code and walk away. This would have been fine with me, but only after IBM had helped to make us successful. We also faced the continuing problem of raising another round of financing. The IBM agreement would be of little use in doing so unless someone there was willing to confirm its existence to prospective investors.

The solution to all this seemed obvious: make a press announcement. Nothing focuses an engineer's attention like being pestered by friends and family to show them the product, and this process begins in earnest when they first read about the product in the paper. Once IBM went public with its support, it would be much harder for them to back away from us. And it would be a lot easier for GO to raise money with a pile of glowing press clippings.

The first step was to agree on a date for the announcement, so Quinlan took the matter up with IBM's PR department. It seemed that the department's main function was to prevent the release of information, not to hasten it. The publicists insisted that the announcement of our relationship not conflict with anything else in their schedule. Given the number of pending announcements — most of them matters such as releasing new models of disk drives and awarding grants to charities — the earliest they could allow was July 18, a long way off. The silver lining was that this would give us time to properly brief the press.

When the PR staff at IBM learned that Quinlan had assured us that Jim Cannavino would make the official announcement, they tried to prevent this from happening. To Quinlan's embarrassment,

Cannavino was suddenly unavailable, leaving Quinlan to handle the event himself. "Cannavino will take calls from the press, however," Quinlan said in consolation.

The next stumbling block was the ritual press release. IBM assigned a young publicist to write a first draft, which he did, but without bothering to contact us first. It was a nightmare, proclaiming that IBM had acquired rights to a new technology from GO that would help bring its own handwriting-recognition research to market more quickly. As it happened, he was authorized to draft the release but not to change it. For a company that grew prosperous on data entry machines, it was odd that IBM restrained its own people from using the delete key.

"If you'd like to discuss this, I'll fly out and we can go over it together," the publicist offered. Instead, I was able to persuade Quinlan to start over with our own draft. Even then, the publicist tried to stick paragraphs from his original version back into the release.

Carol Broadbent, our newly hired director of marketing, who was in charge of corporate communications, worked diligently with GO's PR firm to position the announcement in the best possible light. This was a tricky matter, as we weren't actually announcing a product but only a "relationship" and a "technology," things the press was justifiably skeptical about. Carol shuttled me around the country on an endless publicity tour, filling every available hour with interviews at important publications.

One day in mid-April, Carol stormed into my office with a fresh copy of *PC Week*. "Take a look at this!" She waited, fuming, while I scanned the headlines. Sure enough, there was a front-page story announcing that IBM had made a major investment in GO and had licensed its handwriting-recognition software to us — neither of which was true. *PC Week*, a particularly aggressive publication, must have somehow gotten hold of IBM's early draft of the release. "They should rename it 'PC Leak,'" Carol said.

I called the reporter and politely inquired as to why she hadn't checked with us in advance. "You'd just deny it," she explained. It must have been a slow week.

Carol was frantic, and with good reason: we had sprung a leak long before an announcement — a PR person's worst nightmare. We immediately set up a command center to do damage control.

Our first line of defense was to try to contain the story. Carol sequestered me in an office and lined up calls to everyone we had previously briefed, asking them to continue to respect our embargo. It helped a lot that the published story was so inaccurate.

Emerging from the office late in the afternoon, I walked down the hall to give Carol a final report on my assignment. "OK, that's about it," I said. "Everyone's covered, and they're all going to hold the line." She was visibly relieved.

When I got to my office the next morning, she stormed in again, brandishing some rolled newspapers as though about to smack a dog for soiling the rug. "Read 'em and weep," she said, passing the roll to me at arm's length. Both the *New York Times* and the *Wall Street Journal* had run follow-up stories — without contacting us, and sidestepping their responsibility for accuracy by prefacing the news with the disclaimer "*PC Week* yesterday reported . . ."

"I can't believe this would interest them," I said in disbelief.

"Can't even do their own research," she added. "No one called to check with us. It'll be worse tomorrow."

"Why?"

"It's on the wire." She was referring to the Associated Press, which fed news continuously to its subscribers via teletype. The AP was a key source of material for papers around the country.

Carol was right. The story was picked up by hundreds of local newspapers the next day, often edited to dispense with the qualifier about *PC Week* in order to improve readability.

We were both demoralized. "What should we do next?" I asked.

"You better get out there and make some more news," she said.

As urgent as this seemed, there was the more pressing problem of raising money. We had managed to stretch our cash for a few more months, and with the additional IBM money, had enough to last through June. Given the delay in the announcement, Quinlan had kindly agreed to talk to selected investors on our behalf to confirm the IBM agreement. I pressed John Doerr into service.

"I'll ask our existing investors for their pro rata," he said, "but we'll still need a lead to set the price." He was referring to the common practice in which existing investors take part in later rounds

I link arms with Bill Gates (left) and Mitchell Kapor (right) shortly before GO was founded. Gates and I would later become fierce competitors in pen computing. *(© Ann E. Yow)*

Kapor consults his "to do" list on a Compaq 286 portable while traveling cross-country on his private jet. On a similar trip, earlier in 1987, Kapor mentioned to me how awkward it was to make notes this way, which led us to the idea for a pen computer. *(Courtesy of John Doerr)*

Left: Kevin Doren, cofounder and vice president of engineering at GO, proudly displays a circuit board from the first portable pen computer. Beside him is the deskbound prototype that the portable model replaced. The race was now on to assemble and test the new working units before making a presentation to State Farm Insurance in June 1989.

Below: Robert Carr gives the first public demonstration of GO's Penpoint operating system at the PC Forum in March 1991. Afterward, Microsoft unveiled Pen Windows, a competitive product whose similarities to Penpoint seemed to us to be more than coincidental. (© *Ann E. Yow*)

GO investor and board member John Doerr (right) shows me why he is a "full-service venture capitalist" as we return to San Francisco from a triumphant product announcement in Boston in January 1991. *(Courtesy of John Doerr)*

Stewart Alsop (right), the influential publisher of *PC Letter,* looks somewhat bemused as David Liddle, a member of GO's board of directors, presses his point. *(Courtesy of John Doerr)*

An early pen computer user checks his electronic calendar using Pensoft's Perspective, which resembles Penpoint's "notebook" look. Touching the pen to a name that appears in bold (in this case "Esther Dyson") calls up information about that person. (© *Ann E. Yow*)

After tracking down Norm Vincent of State Farm (far right) at an industry conference, I persuaded him and two associates to step into a bar in the hotel lobby for a quick demonstration of GO's latest handwriting-recognition software. *(Courtesy of John Doerr)*

Mike Quinlan (right) and Kathy Vieth of IBM pose for press pictures with me following the July 1990 announcement that IBM would license GO's operating system.

I turn over the reins of GO Corporation to Bill Campbell (left) at the January 1991 announcement of Penpoint. I was CEO of GO during its developmental stage; Campbell had the experience, connections, and stature to help the company meet future challenges. (© *Doug Menuez/Reportáge*)

The GO executive staff takes a rare break together. Bill Campbell, a master at team building, would arrange lunches and other outings whenever he sensed that the staff was becoming stressed out. Clockwise from left: Kevin Doren, Mike Homer, Robert Carr, Randy Komisar, Campbell, Debbie Biondolillo, Stratton Sclavos.

John Sculley, then the CEO of Apple, takes care of business during a conference in Palm Springs. Sculley nearly undermined GO twice: once by asking IBM to back Apple's Newton project, and later by making a similar proposal to AT&T. (© Ann E. Yow)

Plain-spoken Jim Cannavino, whom I first met in a company trailer at COMDEX, in Las Vegas, made a tempting offer of a partnership with IBM. But his resolve to support GO fell victim to the harsh realities of IBM's waning industry influence and its internal battles over operating-systems standards. (© Ann E. Yow)

Jeff Raikes, Bill Gates, and Lloyd Frink of Microsoft test-drive Pen Windows, Microsoft's response to Penpoint. Frink, who visited GO in 1989, was instrumental in designing the rival pen interface. (© Robbie McClaran)

Left: Esther Dyson interviews Bob Kavner at her PC Forum. Kavner, at that time a group executive at AT&T, arranged the financing of EO, the personal communicator company that grew out of GO. *(© 1993, Fran Solomon)*

Right: Alain Rossmann, the chairman of EO, presents the marketing strategy for EO's personal communicators. Rossmann's company, which was swallowed up by AT&T, soon afterward swallowed up GO. *(© 1993, Fran Solomon)*

My wife, Layne, who is also a computer industry professional, often provided informal advice and moral support. At a surprise fortieth birthday party she threw for me, she hopped on the back of my motorcycle as I drove into my living room to liven things up.

in proportion to their percentage ownership, but only if a new investor is willing to set a price and participate.

I knew what to do. My next call was to Norm Vincent at State Farm. After all, this is how insurers made some of their money, by investing the premiums they collect in advance each year and earning as much as they can on the "float" until they have to pay the money back in claims. Also, State Farm was well aware that the IBM deal was for real.

"Norm, this is Jerry. I just wanted you to know that we are about to raise another round of financing, in case State Farm might be interested in investing."

"I'll call our investment department," he said without hesitation.

If there was one group at State Farm that I expected to be different from the others, it was the investment department. I imagined a fast-paced trading floor in a Wall Street high-rise, with nervous M.B.A.s sitting at banks of computer screens, watching every move of the financial markets. I was wrong. Soon I was back at headquarters in Bloomington, but on a different floor with a different mission. The investment department looked as quiet and reserved as a bank. A single computer display sat unattended in a corner, apparently tracking State Farm's $60 billion in investments. I made a brief presentation, then went to lunch at a country club with Norm Vincent and two investment officers, where they inquired about the asking price.

"Our investors are willing to go at two-fifty per share," I responded. This was true, but perhaps a bit misleading. Kleiner Perkins was willing to support any price that I could persuade a new investor to pay. After due consideration, State Farm had tentatively committed $5 million to help us out — at the $2.50 price, a dramatic markup from 75 cents in the previous round. This placed the company's value at close to $75 million, a high figure for GO's stage of development. But I had my lead, and everyone else was willing to follow.

Then John Doerr arranged for us to visit Andy Grove, the CEO of Intel, as a possible investor. Intel had a near monopoly on the CPUs — the central processing units — that are the heart of DOS- and Windows-based personal computers. An outspoken Hungarian, Grove ran his company with the intensity of a field marshal. I went

to Santa Clara and demonstrated our product to him and several
lieutenants.

Grove focused on the prototype. "What microprocessor does it use?"

"The Intel 286," I responded proudly.

He pursed his lips in disappointment. "Why not the 386?"

"Your salespeople didn't tell us about it in time." This was something that really annoyed me. After convincing us to stick with the
286, Intel started putting up billboards all over the place showing a
sign painter crossing out "286" and writing "386" over it in red paint.
Intel was trying to persuade everyone to upgrade to its latest, greatest product. I had even written a letter of complaint to Intel's VP of
marketing. Fortunately, no one remembered the letter.

"We're always interested in good software for our processors, and
that's why we'd be investing. Are you planning on porting to any
other processor line?" Grove asked. He knew that we were as much
a danger as an opportunity. With a new operating system, we had
the option of switching to a competitor's chips if we wanted to;
then, if we were successful, he would be forced to play catch-up.

"Not at this time," I responded guardedly.

"We'll bet five million — but with the understanding that you'll
upgrade to the 386 as soon as possible and agree that new versions
of your software won't be released on competitors' CPUs before
Intel's."

I looked at John. He was trying to nod his head vigorously in
agreement without being too obvious. I thought Grove's proposition was probably OK, but others were investing without conditions.
"I don't know if we want to saddle ourselves with these restrictions
for the same price that others will pay without special concessions."

"*And just who might you be thinking of porting to?*" Grove asked
with eyebrows raised. John clutched the arms of his chair, worried
that I might blow the deal. "Look, Intel isn't like the other investors.
When we announce that we're backing you, it will really help your
market momentum." This was true, and a reasonable argument.
Besides, I couldn't imagine GO redesigning the pen computer
around another CPU, so I agreed to his conditions.

Grove assigned one of his staff to negotiate the contract, and
we worked up a detailed legal agreement covering the restrictions.
Whenever I sounded hesitant, the negotiator would remind me of

how valuable Intel's public support would be. And when we were about to close the deal, he called me up to request a change. "Jerry, we've just had a meeting about the investment, and we've changed our mind about a few details."

"Like what? I thought we were done."

"We're only willing to invest three million, not five. And we don't want to announce the investment publicly. In fact, we want it in the agreement that you won't tell anyone at all, except for a few investors and other partners. And only under an NDA."

I was quite irritated. "What is this bullshit?"

"This is what we're willing to do. We can revisit the announcement question again after you upgrade your system to the 386."

"I'll call you back." I hung up the phone and called John Doerr.

"John, Intel's trying to jerk us around. They cut the investment to three mil, and won't announce."

"Calm down, calm down," John said. "I think Grove may be in the middle of negotiations with Bill Gates, so he's probably sensitive about rocking the boat right now. It's just good business." He thought for a moment. "We need the money. Let's do it anyway."

He was right. Replacing the $3 million would have been difficult, and the other investors were expecting Intel to participate. Reluctantly, I called back Grove's negotiator and agreed.

With State Farm and Intel spoken for, it was a simple matter to round up the rest of the money. By mid-June we closed on $15.31 million, just in time to pay the bills for the announcement with IBM.

"Sue King's very concerned about your schedules and the recent changes to your user interface." Mike Quinlan sounded as if he were letting me know I had bad breath right before a blind date. "She's recommending that we hold up the announcement."

I called Sue directly to find out what the problem was.

"You added icons to the system without telling us," she said. "Now it looks a lot more like OS/2, so that group is arguing once again to kill the project and go with them instead."

At first I couldn't imagine what she was talking about. Then I realized what it was: after careful study, our user-interface design team had enhanced the "notebook metaphor" to allow for multiple

notebooks. These appeared as tiny pictures of books at the bottom
of the screen. When you wanted to open one, you tapped on it. This
was only superficially related to the icons in Windows and OS/2, but
that was enough to threaten the entire IBM relationship yet again.
I had to think fast.

"Sue, would you please tell the OS/2 people that those aren't
icons? It's a bookshelf."

"What?"

"That's part of the metaphor. It's a bookshelf where the notebooks are kept. Besides, it's across the bottom of the screen, not
along the right side like in OS/2."

She took this argument back to fight off the OS/2 group and
soon dropped the request to delay the announcement.

Finally, the announcement week arrived. IBMers showed up at
GO two days early for the purpose of reviewing what everyone was
going to say. Quinlan brought along a new IBM face, a woman
named Kathy Vieth, who he had previously informed me would be
joining him to help promote the project both inside and outside
IBM. Then Quinlan took me aside to tell me some bad news, which
he had a habit of holding back until the last minute: after the
announcement, he was "retiring" and Kathy would henceforth be
GO's primary contact.

The last thing we needed at that point was another changing of
the guard, particularly since it was clear from watching them that
Kathy Vieth and Sue King didn't get along very well. After working
to develop a relationship with Quinlan, I wasn't looking forward to
starting from scratch again. But the show had to go on.

Despite the premature disclosure in *PC Week,* Carol Broadbent
had managed to pack the room we rented at a nearby hotel with
more than seventy reporters by promising them more news and free
food — I'm not sure which was the more potent draw. After the
crowd picked over the cheese and fruit, the moment for our first
public announcement was at hand.

The staging was spare, like a Shakespearean set, except for a
gleaming silver construction of our newly redesigned logo, illuminated
by a spotlight: two interlocking circles that spelled GO in script,
both right side up and upside down, enclosed in a blue oval. Our

new VP of marketing, recently stolen away from Steve Jobs at Next, had insisted on paying $70,000 to a high-priced design firm for this blue egg. "It's a bargain," the marketing man had said. "Next paid a hundred grand for their logo." He wasn't with the company long.

I was punchy for lack of sleep. As Carol dimmed the lights, I leaned over to Quinlan and whispered, "This is your last chance to back out." Fortunately, he laughed.

I got up and approached the podium to speak. "I'm pleased to have this opportunity to announce to you a new partnership between GO and IBM in the area of pen-based computing." Suddenly the lights went out, strobe lights started flashing, and warbling sirens wailed so loudly that people covered their ears.

The fire alarm had gone off. It was impossible to continue, and the hotel refused to shut off the alarm until the fire department arrived. I stood at the podium for several minutes, making hand signals to the audience to stay put. I considered doing the announcement in pantomime.

Of course, it was a false alarm. By the time the mayhem ceased, the audience was thoroughly distracted and partly deaf. I needed to regain their attention. "I knew this announcement was hot, but not that hot!" The room exploded in laughter. A number of reporters used the incident as the lead for their articles.

When I was through, Quinlan gave a convincing speech about IBM's unwavering commitment. Then Kathy Vieth got up and seconded Quinlan's remarks.

After the event, I instructed my executive assistant to send Quinlan an appropriate gift. Quinlan called me up the next day. "Thanks for the beautiful bouquet," he said.

"Oh shit, is that what she sent?"

"Don't worry, my wife isn't the jealous type."

Despite the earlier leak, the press was kind to us, giving us front-page coverage in most computer publications. Except for *PC Week* — they buried the story on page 135.

▣

The press had been unfair to John Sculley at that time, portraying him as a manager with little vision, a mere caretaker for Steve Jobs's

Macintosh legacy. To improve his image, he was looking around at Apple for an exciting technology that he could take ownership of. He picked the pen computer project, by now code-named Newton, which had languished since Steve Sakoman's departure. He asked Larry Tesler, his technology guru, to take charge.

Sakoman, a hardware type, had focused much of his attention on a secret joint-development project with AT&T's Bell Labs. AT&T had custom designed to his specifications a hot new type of low-power but high-performance microprocessor known as a RISC chip, called the Hobbit. There was only one other similar chip available, called the ARM, for Acorn RISC Machine, offered by a British technology company.

Tesler began his tenure with a thorough housecleaning. He asked AT&T to make changes in the Hobbit's design and to produce a delivery schedule and price list. Finding AT&T's response unsatisfactory, he canceled the Hobbit project and switched over to the ARM, to the horror of Bell Labs engineers and even some of his own staff.

Then Tesler did the unthinkable. At Sculley's request he considered canceling Apple's own software project and working with GO instead. The idea of buying technology rather than developing it internally was entirely contrary to Apple's culture. This technical xenophobia was known in Silicon Valley by the initials NIH — not invented here. NIH must have originated at Apple, because by nature the company was incapable of adopting the term from outside.

To my delight, Tesler had hired a consultant to advise him on the future of the Newton project. It was Peter Miller, the person whom Mitchell Kapor had originally asked to join *his* team, and who by that time was a personal investor in GO. I helped things along with an unbeatable financial proposal for the licensing and control of our software, similar to the IBM arrangement.

But by the time Peter delivered his report, Tesler was facing a growing revolt on the Newton team. He sympathized with their desire to maintain internal control over the technology and identify this project solely with Apple. The prospect of buying a key component from the outside — particularly one that IBM had beaten them to — wasn't very palatable, and the team also had concerns about the suitability of the GO technology. So Tesler recommended to Sculley that they pass on working with GO. Sculley reluctantly concurred, and the Apple team found themselves starting anew with a fresh mandate from management.

Although it seemed impossible, the pace of activity at GO continued to increase. There were so many urgent matters demanding my attention that I took to staying at a nearby hotel, to save the few critical minutes spent commuting from the city. There was no such thing as a weekend off.

Engineering schedules were slipping again, so we had a round of meetings to decide how we could remove certain features from the final product. Sue King continued to schedule review meetings. The last stock-purchase agreements permitted us to raise an additional $4 million within ninety days of the most recent investment, so I hosted an unending stream of potential investors. Following the success of the IBM announcement, many developers wanted to begin writing applications for our operating system, but most of them insisted on meeting with me first. I was needed for press briefings, recruiting key employees, performance reviews, reports to investors, and, most of all, for visits to other personal computer manufacturers that might follow IBM's lead in licensing our software, many of whom were in Japan. Despite my constant efforts, I was falling behind. I called John Doerr for advice.

"John, I'm burning myself up. I'm working the equivalent of two full-time jobs. What should I do?"

"That's easy, let's split your job in half." The logic of this simple observation was compelling. "Engage a headhunter and we'll look for a chief operating officer. And get out and do something this weekend." I took both suggestions.

Coincidentally, I had been invited to a dinner party at the apartment of a woman recently hired in our sales department who lived a few blocks from my condo. I accepted. After the meal, the guests decided to take a walk. As it happened, the woman had a friend who was giving a party nearby, on the way back to my house. The two of us trekked up Russian Hill — no small feat for someone with little sleep who hadn't worked out for months. By the time we arrived at the top of the hill, I was out of breath. She pointed to a five-story walkup.

"What floor?" I asked.

"Five," she said.

I shook my head. "No way. No way. I'm going home."

"She's got lots of female friends . . ."

"If I'm going to climb those stairs, I damn well better meet my wife at the top."

We went on up. After catching my breath and checking out the desserts, I scanned the room with the practiced eye of a thirty-eight-year-old bachelor. My rusty radar settled on the host, a bubbly, slender woman named Layne. She had sparkling blue eyes and long blond hair pulled back in a ponytail reminiscent of a cheerleader, which in fact she once was. She worked in marketing at a PC company.

I introduced myself. Layne, in turn, introduced me to her boyfriend, an athletic-looking lawyer who worked at Next. As I left, I slipped her my card. "I know I look like a nerd," I said, "but I'm only a partial nerd. If you ever dump that hunk, give me a call."

Three months later, she did. The next August, Layne and I would be married.

"Seven thousand!"

The crowd was going wild. I had donated a one-hour piano recital — to be performed by me, in my condo on Lombard Street — to a charity auction for the Foundation for Educational Software. It was one of many crazy items, including a software code review with Bill Gates and an appearance as a character in a video game. The event was staged one balmy evening around the pool at a resort in Laguna Niguel during Agenda, the exclusive invitation-only conference held each September by the computer industry analyst Stewart Alsop. The top four hundred movers and shakers in the business were there.

Steve Ballmer of Microsoft was the auctioneer, and he had whipped up the audience after a fabulous dinner of chateaubriand and cabernet sauvignon, hosted by an investment-banking firm. The items were going for staggering sums, a testament to the fortunes made around personal computers. I had boldly estimated the value of my contribution at $250.

"*Eight* thousand!" someone else yelled out.

Ballmer pounded his gavel. "Sold."

"I better start taking piano lessons!" I yelled across the pool to the lucky winner, who slapped his forehead in mock disgust. The crowd loved it.

By this point, the guests were on their feet. John Doerr was on one side of me, Vinod Khosla was on the other. Vinod leaned over and spoke into my ear. "I hear that Bill Campbell might be loose in the socket at Apple." Campbell, who had hugged Steve Sakoman in an Apple parking lot when I first met John Sculley, was the CEO of Claris, the applications software division of Apple Computer that Sculley had decided to spin off as a separate company three years earlier. But when Claris was ready for an IPO, Sculley decided that Apple should be in the applications business after all, so he killed the plan and folded Claris back into Apple. Campbell hadn't made a secret of his displeasure.

"Do you think we have a shot at him?" I asked. Campbell had a reputation as a charismatic leader and an outstanding manager, but seemed more like a big-company guy.

"We'd have to make him CEO, if that's OK with you," Vinod yelled over the noise of the loudspeakers.

I thought for a minute. I wasn't in this for the job title. In fact, I rather disliked the responsibility of managing a growing staff, which now numbered more than a hundred. Besides, we hadn't turned up any good candidates for chief operating officer, despite many months of interviewing. "No problem — if that's what it takes to attract someone of his stature."

"You can have the position of chairman," John said.

"Sounds good to me."

"Sixteen thousand!" Ballmer called out. "Come on, everyone, this is the last item!"

"What's the item?" I asked John. Instead of answering, he grabbed my hand and stuck it into the air, holding my bid card aloft.

"Seventeen thousand from Jerry Kaplan!" Ballmer was going hoarse.

"Oh, hell. What's the item?" I asked again.

"Sold! A two-hour business plan consultation with John Sculley. Good luck, Jerry!"

I was stuck. Vinod and John were laughing so hard they could barely breathe.

As the evening ended, someone asked me what I planned to do with the time. "Break it up into five-minute segments," I said, "and sell them to everyone who wants to tell Sculley off." I was telling this to the wrong person, a widely read newspaper columnist. To my embarrassment, he published my remark the next day.

□

By late October, a rumor was circulating in the company kitchen that our engineering schedules were going to be delayed again, possibly into 1991. This had become the midlevel technical managers' standard technique of gauging the reaction to a schedule slip before actually proposing it. GO was rapidly gaining a reputation as a purveyor of "vaporware," the industry's term for products that are promised but never delivered. I had to take some action or we were never going to get a product out. So I called a meeting of the relevant managers: product marketing, quality assurance, documentation, operating systems, applications, technical support, hardware engineering, manufacturing, and public relations.

"It's been three months since the IBM announcement, and all I hear on the phone is 'When are we going to see a product?' We can't hold out on nothing but promises much longer. It's time to set a product launch date." I could see everyone freeze. "Carol, what do you think?" I already knew what she thought. She spent much of her time fielding calls from journalists who teased her with information about our potential competitors.

Her voice was strained and tense. "We just got word that Microsoft is going to announce in February that they're planning some pen extensions to Windows. We can't let them steal our thunder. We have to announce in January."

I was hoping to build a consensus, but instead a revolt broke out. The operating system wouldn't have enough test time. The application developers' documentation, which had grown like a cancer to several thousand pages, wouldn't be finished. There was no guarantee of how many demonstration machines we could manufacture by then. Only some of the promised applications would be complete. User testing indicated that the pen computer wasn't yet ready for sale to end-user customers, only for applications development.

"I'm glad we're all in agreement," I said. A nervous laugh rolled around the room. "Look, gang, it's time to bet the company. IBM is getting antsy again; Microsoft is breathing down our neck; the investors are anxious; and our reputation is fading. We can't afford to delay any further, period. It's time to stand and deliver."

Carol interrupted. "I can get the Meridien Hotel in San Francisco for January 22, but only if we make a fifty-thousand-dollar deposit now. Then we can take a redeye to Boston, where the Boston Computer Society will let us present at their January 23 meeting."

I looked around the room. Everyone's face was white with fear, wondering if I had the nerve to make this substantial deposit to reserve the hotel. "OK, Carol, book it. Listen people, figure out what you can get done by that date. This is it — we're either going to make fools of ourselves or take the high ground in the coming debate." The room was silent.

"I'll do my part." It was Phil Ydens, the stoic operating systems engineer. "Whatever it takes, my group will be ready."

"Same here," said Celeste Baranski, the director of hardware. "Tell me how many units you want, and we'll deliver."

"Maybe we can quick-print copies of the key manuals for the launch. That'll save us several weeks," said John Zussman, the director of documentation.

It was working. With a fixed endpoint in sight, the team was pulling together. After some discussion, we agreed to pitch the product as a "developer's release" — an early version aimed at applications developers — and to call the event a "developer's summit." There was one last thing to do.

"It's time to name the product. Carol, show everyone our proposal." She pulled out a large piece of foam-core board with our new product name and logo on it.

"Penpoint," it said in large black letters. A line ran across from the *e* to the *t*, and an inverted red triangle hovered above the *i*, dotting it like the point of a pen. Underneath was a tag line: "The Pen Is the Point." Everyone loved it. We hung it up in the kitchen as a battle standard.

"Just one thing to watch out for when you demo," Carol said. "As you know, the handwriting recognition tends to run words together. Be sure and space out your words when you write the tag line, or you'll wind up describing male anatomy."

When word spread inside IBM that we had committed ourselves to an announcement date, it was open season on renegotiating our agreements. Out of the blue, I got a call from John Kalb, the IBM negotiator who had disappeared after my last meeting with Jim Cannavino. "We never completed the agreement to share in your revenues," Kalb said.

"What agreement?" I protested. "Quinlan and I worked out a completely different arrangement!"

"That's not Cannavino's view. We simply haven't finished the negotiation. He wants twenty-five percent of your revenue from others as an offset against our royalty payments."

I was shocked. When I met with Cannavino in Armonk, I had proposed giving IBM a credit against royalties for a portion of our revenue from others — but only in conjunction with IBM's buying 25 percent of GO at a high price.

"No way, John," I said.

"Not only that," Kalb said. "If the credit exceeds what we owe you, we get the difference in cash."

"I want to talk to Cannavino."

"That isn't possible. Until you agree, there will be no more public or private support for GO."

Later, I asked David Liddle to check this out with Cannavino. He did, and was amazed to confirm that Cannavino was quite serious.

We were screwed. And now Kathy Vieth, who was supposed to be our emissary on such matters, was unavailable, despite my understanding that she would call a number of potential Penpoint licensees and applications developers who were eagerly waiting to hear from IBM. That night I discussed the situation with Layne, who had an uncanny knack for guessing people's intentions. "Don't kid yourself, this is no coincidence," she said. "Kathy's been told to disappear, to put pressure on you. Remember, when push comes to shove, her allegiance is to IBM, not GO."

I called a board meeting to deal with this emergency.

"You're going to look pretty silly standing up there in January without IBM," said John Doerr.

"In other words, we have a choice," I said. "Go out of business now, or go out of business later."

"Sometimes a rat has to gnaw off its leg in order to live," David Liddle added.

"Sure," I replied, "but rats don't gnaw off their legs when they're dangling by them!"

Vinod Khosla summed it up. "You have to make a deal one way or the other."

I had only one piece of leverage, my rights under the "joint walk-away" provision in the license agreement. I knew IBM was committed to State Farm, and by now several other large customers as well. I called Kalb.

"John, as it stands right now, we have the right to cancel our contract with you and not deliver any further versions." He checked the contract, and was surprised to discover that I was right. "So here's what I propose. We'll give you twenty percent of our revenue on a sliding scale, but only after we have income of fifty million dollars on Penpoint. And it's only a credit against your royalties — no cash. In return, I'll cancel GO's right to walk away from the relationship." After several weeks of further discussion, he consented.

These negotiations went right down to the wire: IBM was still threatening to back out a week before the January announcement. We had managed to avert the crisis, but I secretly knew that there was only one way for us to survive, and that was if IBM failed in the market, leaving its royalty credits unused. IBM had inadvertently made it in our interest to decrease its market share at the expense of other licensees, who would pay us real money in royalties.

After completing final press briefings for the dailies — the local newspapers plus the *New York Times*, the *Wall Street Journal*, and *USA Today* — I rushed over to the Meridien Hotel to check on the preparations. We were allowed access to the facilities only one day in advance of the product announcement, so Carol Broadbent had to lay her plans with the care of a police chief about to storm an orphanage held hostage by terrorists.

The cavernous main ballroom was filled with teams of specialists, each moving feverishly to complete their assigned tasks. Construction crews worked everywhere, rigging giant projection screens, hanging Penpoint pennants, planting row after row of chairs, and building a custom stage set. Following close behind were audio-visual technicians, securing wires with yellow-striped safety tape and adjusting the sound levels of numerous wireless lapel microphones, each labeled with the name of a speaker. The cramped area behind the stage looked like the control room of a nuclear power plant: pillars of equipment were set to project videos, live images and slides, and provide synchronized music. Stagehands reviewed minute-by-minute scripts in small groups, practicing their cues and sequences. Overseeing the fray was a director, in constant communication with each crew via a wireless headset, like an electronic coxswain.

But this was only one of a dozen rooms being prepped for the event. The two-hour announcement was to take place in the morning, then, after a buffet lunch, more detailed technical presentations were scheduled throughout the afternoon in the hotel's smaller conference rooms. Maps and signs hung everywhere, directing foot traffic to these "breakout sessions."

One room was marked "No Admittance: Charge and Load Station." Inside, the GO buildmasters — who had taken the program modules submitted by the programmers and assembled them into a software "build" — were cooking up their final system and loading it into sixty-five spanking-new, sleek, black pen computers, wrapped in leather cases similar to my original maroon portfolio. The product-marketing crew had created an electronic notebook full of wonderful goodies, from traffic-accident forms to architectural drawings to memos recommending new recipes.

The rehearsals began at three, and I was up first. The director showed me my mark and video cue, then motioned me to the podium. I started practicing my speech. "Look directly ahead more, so our camera unit at the back of the room can pick up your eyes," he said. When I finished, Robert went through the paces for his demo. Next up were presentations by the ISVs, the independent software vendors.

We had invited two of the best to present: Slate Corporation, a collection of PC software industry luminaries headed by Vern Raburn, a veteran executive of both Microsoft and Lotus; and Pensoft, a group of relatively junior but talented engineers headed by a young entrepreneur named Michael Baum. About six months earlier, I had made a deal with Raburn: GO would stop development on our own electronic forms application and turn the project over to Slate if it would agree to finish the application and bring it to market under Penpoint. Now Raburn was ready to show off Slate's product, called PenApps, for the first time. Carol and I had repeatedly asked for an advance copy of his demo script, but because of unexplained delays, it somehow never showed up. Pensoft had been working on a flexible, pen-based "personal information manager," similar in some respects to Lotus Agenda, which would track handwritten appointments, notes, and phone lists.

Raburn got up to practice his presentation. To my surprise, he had directed his engineers to build a calendar and appointment scheduler with PenApps, along with two other unrelated sample uses, which he planned to demonstrate. When he climbed down from the stage, he avoided my eyes.

"Vern, let's talk," I said. "You know what Pensoft is doing. We're here to build a market at this point, not snipe at each other. If you want to shoot at him later, that's your business, but this is GO's announcement, and I want you to stick to the electronic forms."

"These aren't applications for sale, just examples. Anyway, I have to pursue my business the way I see fit," he said.

I felt like inviting him outside for a fistfight, but there were more important matters to deal with. "How about if you stick to the calendar and skip the appointments," I told him. "And I want you to lead with a heavy emphasis on the forms."

He reluctantly agreed.

Completing the roster, along with IBM, were senior executives from NCR and Grid Corporation, who had decided, after an extensive courtship, to follow IBM's lead and license Penpoint for their future pen computer projects. In the afternoon sessions, forty other ISV and development partners would be demonstrating or discussing their applications.

Near the end of the rehearsal, Carol walked up to me and said, "We've got a problem." She seemed clinically devoid of emotion, a defensive technique she had developed to deal with the unimaginable stress of pulling this all together on schedule. "The cables that run the images out of the live demo units from the podium to the display screens aren't working. Can you please talk to Kevin?"

I found Kevin in a corner of the room, sweating over a pile of electronics with a soldering iron, attended by a circle of his engineers. "My guess is that the signal degrades over the long run of the cables," he said. "This worked fine with the shorter lines in our tests."

"So now what?" I asked.

"If I'm right, there's no easy fix. We need to find a way to boost the signal."

"Maybe we can rig a camera overhead, aimed down at the podium."

"Might work, might not," Kevin said without looking up. Carol got the video crews to work through the night to set this up.

The next morning, I drove back to the hotel. I knew that within a few hours, either we would be the laughingstock of the computer industry — having blown $25 million in a foolhardy attempt to convince people that pen computers were something new and different — or we would be hailed as visionaries. I still did not know which way this $500,000 show was going to go.

Carol gave the signal to open the doors, and seven hundred invited guests flowed into the ballroom. The transformation from the previous evening was dramatic — everything was orderly, polished, and clean. With broad banners adorning the walls, the room had the feel of a royal pageant. At precisely ten o'clock, the lights dimmed and a fast-paced video rolled on the center screen above the stage, accompanied by loud rock music. An announcer's voice said, "And now, Jerry Kaplan, CEO of GO Corporation." I walked to the spotlighted podium to begin my speech.

"It's my pleasure to welcome you here on the birthday of the man who made penmanship famous — John Hancock." The crowd burst into laughter. "We've asked you here today to witness a demonstration of one compelling truth, that computers operated by a pen will be simpler and more convenient than any other computing tool for people on the go. Adding a pen to a system designed for a keyboard makes it more complex, not simpler. The pen is not an option, *the pen is the point*." The audience applauded enthusiastically. "It's my great honor to introduce to you the first operating system designed specifically for the coming wave of pen-based computers." After a brief summary of Penpoint's key features, I introduced Robert.

He walked to center stage with his demo unit unobtrusively tucked under his arm. Unsnapping the leather cover, he pulled out the inch-thick, four-pound tablet computer, with only a screen and a single on-off button, and held it aloft. "Here it is, the future of computing." The audience cheered again and began stomping their feet. The GO engineers among them started to chant "Pen-point, Pen-point, Pen-point . . ." Soon the whole crowd was into it — it sounded like a wartime rally. Robert plugged his computer into the display cable and looked over to the side of the stage at Kevin, who had been up much of the night trying to get it working. Kevin pressed a button and all three screens flashed to life with sharp images of an electronic table of contents. He shot Robert a thumbs-up. Carol, also in the wings, gave Kevin a big hug. The director's face broke into a broad smile. Then Robert launched into his demo.

He began with a quick tour of the notebook metaphor. Each document occupied a page, identified by a page number in the upper right-hand corner. Labeled tabs ran down the right side of the screen as in a loose-leaf binder. After turning to a memo, he demonstrated Penpoint's "gestures," scratching out some words and watching as the surrounding text closed up. He drew an insert caret, and a combed box called a writing pad popped up. After writing a few words, one letter to a comb, he tapped a button labeled "OK." The new handwritten words were instantly transformed into typed text and inserted into the memo. Our handwriting team had made great progress, and it showed — though there was still a long way to go before the software was good enough for general use.

Then Robert turned to a clean page, where he drew a circle freehand. When he raised his pen, the circle immediately snapped to a perfectly round shape. He drew an X over the circle, and it disappeared. He sketched several boxes, which melted into perfect rectangles, and dragged them with the pen into the form of an organization chart. He then connected the boxes with lines, which adjusted themselves to be straight and level, and wrote the word "simple" over one of the boxes. The translated word appeared as text inside the box. He turned to a page containing a faxed document and showed how it could be marked up with the pen and faxed back. Finally, he demonstrated "math paper," which looked like regular lined paper, but when he wrote the mathematical expression "$2 \times (3 + 4)$" on a line, it added the answer "$= 14$."

After the other speakers were done, I returned to the podium to wrap things up. "I'd like to introduce you to the latest addition to our management team. As of today, we have a new president and CEO. Please welcome Bill Campbell!" The audience cheered as this renowned executive took the stage. He gave a brief but inspirational speech about GO's vision and commitment to building a new market.

When we broke for lunch, forty GO staffers roamed the crowd with pen computers to give people a hands-on feel for the new technology. And to our surprise, nearly everybody stayed for the afternoon sessions.

In the evening, we held a company party at the hotel. The mood was electric. Bill Campbell announced that we were granting every employee a bonus of 286 shares of stock — a reference to the Intel processor powering the demo units. For a while I played the piano, and someone put up a jar for tips, which quickly filled up. Then several of us were off to the airport for the redeye to Boston.

Although we were exhausted, we spent much of the flight showing off Penpoint to interested passengers. The next day, at the Berklee Performance Center in Boston, we repeated the San Francisco performance to an even larger crowd, eight hundred members of the Boston Computer Society.

When the dust settled, the press had delivered our message loud and clear. We were the cover story for *Byte* magazine and *PC World*. Stewart Alsop wrote us up in his next newsletter: "GO has developed what is, in my opinion, the most state-of-the-art personal computer operating system in the industry." We were inundated with calls requesting application development kits and asking where our stock could be purchased.

I took Robert, Kevin, and Bill out for a celebratory drink. "Well, gentlemen," I said. "I think we made our point."

"No question," Robert said. "We're riding the top of the wave."

"A fuckin' tidal wave, judging from the press," Bill added.

Kevin raised his glass and said, "Now we just have to keep from drowning in our own storm."

9

The War

BY THE BEGINNING OF 1991, Microsoft had become the most powerful player in the PC industry. Founded in 1975, the company had spent its first decade getting less respect than it deserved, before quietly blossoming into an international force. Its image problem stemmed from the fact that no one on the outside could discern the company's business strategy. Most successful companies select one market or a few related markets, develop expertise in those areas, then build a defensible position with quality products or services. Microsoft, however, sold a wide variety of seemingly unrelated software products in a number of different markets. This omnivorous behavior was the result of a unique approach to survival that might be described as selective opportunism.

Rather than focusing on any one thing, the company became adept at identifying promising market niches with weaker competitors. It would closely study their products and tactics, then launch an attack on their position with a strong product and aggressive pricing. Sometimes Microsoft would propose some form of cooperation or joint development, to learn about the market before staging its own entry. This was the corporate version of the cheetah's hunting technique: keep a close eye on your prey, sneak up, then outrun it. Like the cheetah, Microsoft became well adapted to its game: lean and mean, fast on its feet, observant, shrewd. And the first warning of its presence was often an unexplained rustling in the tall grass.

"I've been thinking about Windows, and I'm having second thoughts about writing applications for Penpoint." One of our independent software developers had called me to discuss the potential market for Penpoint applications. "Your system is compelling, but Windows is compatible with an installed base of millions of PCs. As a developer, it's hard to ignore that."

The afterglow of our announcement had begun to dim, and Microsoft evidently decided it was time to strike before we gained too much momentum. We had heard rumors for months that its engineers were working feverishly to add some pen extensions to the increasingly popular Windows interface, but it seemed that the runaway success of our event may have taken them by surprise. Microsoft arranged a private conference with key software developers to reveal its plans, taking particular care to target those who had publicly committed themselves to Penpoint. The race was on.

"Windows may be widespread, but none of those computers have pens attached," I said. "Writing with a mouse would be like writing with a potato." I didn't know exactly what Gates's people had done, but it seemed logical that they had simply rigged a pen as an alternative to a mouse.

The developer still had his doubts. "Sure, but it's going to be a lot easier for people with Windows apps to modify their existing programs than to start fresh with Penpoint, no matter how much better it is."

I could tell by his voice that, like many independent developers, he wanted to know that he was really appreciated. "You don't want to do just another Windows spreadsheet, just another Windows word processor. Where's the sport in that? You're a creative guy, and you need a system that will let you build something truly different. Windows is a crowded market, Penpoint is a wide-open playing field."

He liked my tack. "I'll think about it and call you back."

"My door's always open if you want to bat around new ideas."

When I hung up the phone, I realized that there was no way I could personally persuade every developer this way. If Microsoft could introduce doubts into one developer's mind, it was surely undermining our efforts with others as well. We had to devise a strategy to keep our story fresh and attractive.

The first step was to find out as soon as possible just what Microsoft had cooked up. It didn't take long.

I spent what little free time I could find during the month of February cooking up a special surprise for my girlfriend, Layne. It was nearly a year since I had struggled up the stairs to the party at her house, and I felt it was time to get serious about our relationship. So I engaged an artist and a programmer to build a special Penpoint application. Then I picked an ordinary weeknight and asked her to come to my house for dinner after work.

"Would you mind doing a reality check on these figures?" she said, handing me a spreadsheet she was working on. Then she noticed I was opening a bottle of chardonnay. It was unusual enough for me to offer her a drink during the week, but she became especially curious when she saw that it was a forty-dollar bottle of wine. "Hmm, Far Niente," she said.

She sat down on the couch and I pulled out a GO prototype computer from my briefcase. "I'd like you to try out a new application," I said. She looked askance at me and pressed the on button.

The screen came to life with what looked like the cover of a comic book. Across the top it read "The Cheerleader and the Nerd: An Interactive Hyperstory." The screen was divided down the middle, with a drawing of a pompom girl on the left and a programmer at his computer on the right. The characters' features were clearly recognizable. "Tap on the people with the pen," I explained.

When she touched the pen to the cheerleader, a cartoon balloon appeared above her head: "Will I ever find Mr. Right?" When she touched the nerd, his balloon said, "Will I find leftover eggrolls in the fridge?" Layne giggled with delight.

The next twelve screens parodied scenes from our time together, painting a whimsical picture of how our lives had grown intertwined. One frame was titled "Layne Takes Jerry to the Opera." The nerd said, "Jeez, this is a lot longer than *Tommy!*" The cheerleader responded, "Psst! I brought your Nintendo Gameboy!"

Another screen showed us mooning over a computer, combining our hard disks. "We'll use COM1 for your mouse and COM2 for my modem," said the nerd. "We'll run DOS from one partition and Unix from another," said the cheerleader.

Still another screen was labeled "Layne Teaches Jerry about Xmas," in which we were decorating a tree. My character said, "Uh, Christ died 2,000 years ago, so we put a dead tree in the living room?" Hers explained, "It's like an IPO for peace on earth and good will to all."

Finally there was a drawing of me saying, "She's really wonderful! How can I show her how much I care? All I know about is computers." In the next screen I'm on one knee, holding the GO pen computer toward her. The caption reads, "Layne, will you marry me? Please write your answer here on the computer."

The real Layne took the electronic pen in her hand and with tears in her eyes wrote "yes" in the space indicated on the screen.

The last panel showed the two of us kissing, surrounded by hearts and birds. My character's left hand was pointing up to the top of the computer, toward the slot where a modem could be inserted.

Layne opened the slot. Inside was an engagement ring.

▣

As a welcome counterpoint to the huge scale of COMDEX, there are a number of smaller computer conferences that attract particular segments of the industry. One of these is the PC Forum, put on each spring by Esther Dyson for five hundred or so of her closest friends and subscribers to her widely read trade publication, *Release 1.0.*

Esther is the eldest daughter of the renowned physicist Freeman Dyson, from whom she inherited great intelligence and an inclination toward the eccentric. After graduating from Harvard, Esther became a journalist and Wall Street analyst, eventually apprenticing herself to the publisher of an electronics industry newsletter. In short order the publisher decided to quit so he could focus his efforts on the venture capital business, and sold the newsletter to Esther. This person was Ben Rosen, who later backed Mitchell Kapor at Lotus.

Short and unassuming, with the cropped brown hair of an Olympic swimmer, Esther eschews the usual female vanities of makeup and fancy jewelry. A warm and wonderful woman, she is unceasingly loyal and thoughtful toward the wide circle of professional friends she has made.

Despite the best efforts of her friends, Esther never developed a normal home life. To this day, she lives like a sort of modern nomad, traveling around the world in a perpetual quest to feed her insatiable intellectual curiosity. For the past several years, she has taken an interest in the former eastern bloc countries, where she has befriended computer entrepreneurs throughout Russia and Poland. She arranges for several of them to attend her conferences, where they follow her around like baby chicks behind a mother hen.

As the theme of her 1991 conference, Esther chose "Beyond the Desktop: Networks, Notepads, and Legacies." She rented an entire resort in the foothills outside Tucson, Arizona, to accommodate the attendees. By mid-February, she was calling around for final confirmations of her speakers. Being a regular on her panels, I was one of the last ones she called to discuss the content of my talk.

"Jerry," Esther said, "I would like you and Robert to show Penpoint to the whole crowd this year."

"No problem. Thanks for the opportunity."

"But I want you to know that I've also asked Microsoft to demonstrate. You can go first if you want."

"That'd be great." I couldn't wait to see what they had.

Even during the conference, Esther likes to get up at six A.M. and swim for an hour while everyone else struggles out of bed in time for the start of the eight o'clock morning session. Normally, people linger over the well-stocked breakfast buffet just outside the main hall. But on the morning of my presentation, the hall was packed in anticipation of the dueling demos. In order to get a good look, several people had dispensed with decorum and sat down on the floor between the stage and the first row of tables.

The gooseneck microphone squawked loudly when she pulled it toward her mouth. She introduced me, and I went to the podium.

"I'm going to be brief, because the best thing for you to do is see our system in action. We're here to answer just one question. Why a new operating environment like Penpoint? Let me state the obvious. Portable pen computers are different and will appeal to a new class of users. Penpoint delivers a radically simpler, more compelling interface that doesn't inherit the complexities of desktop systems. I want to turn over the session to Robert Carr, our vice president of software, who's going to give you a demonstration."

Robert walked to a table on the stage, where he plugged his computer into a cord that led to a large display above his head. By this point, the floor in front of him was crowded with people sitting cross-legged. "This looks like kindergarten," he said. "I want you all to hold hands. We're going to sing 'Kumbaya.'" The crowd broke into laughter, and Robert launched into his demo.

He was brilliant, as usual. He walked the audience through the notebook user interface, explaining how each document appeared on a page. Then he filled out a "property appraisal form," which included an area for sketching a building lot, and added a signature at the bottom. He showed an architect's floor plan for an office, in which he moved a table by tapping on it with the pen and dragging it to a new location.

He concluded by explaining the difficulties of creating a generic "layer" on top of a keyboard-and-mouse application which could do these sorts of things — that to tap the true potential of the pen market, a fresh approach was required. To demonstrate his point, he drew three circles and showed how Penpoint interpreted each one. The first circle was in an open space on the screen, where it snapped instantly into a perfectly round shape. Next he circled a word, and it popped into an "edit pad." Then he wrote the letter *o* in an area where it was translated into typed text and inserted into the word. "The same shape — three times in twenty seconds — and they have three very different meanings. They are all intuitive to the user and are only possible if the applications are rewritten." The crowd applauded enthusiastically as we collected our gear.

Esther returned to the podium and invited the speakers from Microsoft to the stage. Robert and I sat down on the floor in front, ready to take notes. Robert was fumbling in the semidarkness to find a blank page in his notebook as I watched the next speakers set up.

"Robert," I whispered. "It looks like Jeff Raikes, the guy who signed the confidential cooperation agreement with us, is going to make the presentation."

He looked up, then shook his head. "But that's not the worst of it. Look who's doing the demo."

Sitting in front of us, about to demonstrate Microsoft's competitive entry into the pen-computing market, was Lloyd Frink, the aggressive young engineer who had visited us under the cooperation agreement to study our work in detail.

"So much for their Chinese wall," I said.

"They didn't need to breach the Chinese wall. It looks like they just went ahead and did the work in the applications division," Robert said. He was right. Just because we thought of Penpoint as operating systems software didn't mean Microsoft couldn't think of its competitive response as an application.

I looked around the room and noticed that Bill Gates was sitting at a table several rows behind us. He rocked back and forth with eyes narrowed, intensely focused, like a lead wolf watching his pack stalk some hapless prey.

Raikes began his talk. "I'm a vice president at Microsoft. Given all you have read in the press about the FTC and about the perception of Microsoft as dominating the industry, you might think that being a vice president at Microsoft is like being a field general for the Evil Empire, but I don't really think that's the case." Raikes was referring to recent reports that Microsoft was being investigated for monopolistic practices in the PC market by the Federal Trade Commission. Despite this tepid disclaimer, and after complimentary remarks about Robert's demo, he chose his point of assault.

"I'd like to find out where the audience is positioned relative to investing in Windows today. If you're in the hardware business . . . or are building peripherals . . . if you're a software developer . . . in the sales channels . . . raise your hands if you're involved with one or more of those Windows markets." He looked around the room at a sea of hands. "Basically the whole audience. OK. There is a large investment right now in standard Windows, and this is an opportunity to leverage the investment that you're already making. At Microsoft we had to decide whether to start fresh and do a whole new operating environment or to leverage an existing operating environment. In reading Stewart Alsop's newsletter, I noticed that if we had done this [created a new operating system], it would have been the eighteenth PC operating environment, which I think is also a problem. We found that we could build on the success of Windows."

I leaned over to Robert. "He's trying to position us as just another PC operating system."

Then Frink launched into the demo. He brought up an application that he called Notebook, which looked extremely similar to an early version of our interface.

Robert was trying to stay cool. He whispered through clenched teeth, "They ripped off our stuff!"

But this was only the beginning. Microsoft had used many of the same "gestures" as Penpoint, but they were just different enough to avoid a copyright-infringement suit. Raikes narrated a story about an insurance agent while Frink made an entry in a calendar, in electronic ink, before filling out a form. Then they demonstrated that you could draw a circle or a square and have it snap into a seamless shape — a concept we had discussed with Frink when he visited GO years before.

Robert was livid. "I'm just glad we showed Lloyd our ideas so long ago, so he could only copy our earlier stuff!"

Raikes finished up by returning to the theme of leveraging existing investments in Windows. "We have twenty-one [hardware vendors] that have announced their support for Pen Windows. That's going to mean a lot of sockets for people to plug their work into. We expect to have *kanji,* French, German, Italian available." He glanced approvingly at Gates as he spoke. "OK, you've had a chance to see why Pen Windows can be a great platform for compelling applications, and you've had the opportunity to rethink compatibility and how it can be leveraged without compromise to the user. Thank you very much."

Raikes was executing the classic IBM strategy of introducing fear, uncertainty, and doubt into the minds of its regular customers, causing them to wait and see before committing themselves to a competitor's product.

The audience applauded politely. They knew that the incredible similarity between our demo and Microsoft's couldn't be a coincidence. It had to be a premeditated attack. But virtually all of them depended on Microsoft in some way for their business, and they weren't about to endanger themselves by being the first ones to cry foul. I later learned that this problem even affected the FTC, which had trouble finding people willing to speak out against Microsoft's tactics for fear of retaliation. After the show, in private, some colleagues were more open about their feelings.

"You got fucked, plain and simple," said Michael Baum of Pensoft. "They just figure you don't have the wherewithal to go after them."

"Why would they think that?" I was putting up a strong front, but we both knew why. Corporate law is designed for large companies — it protects only those with deep enough pockets to finance a protracted legal battle. Compared with Microsoft, GO could barely afford the carfare to the courthouse. A lawsuit could tie us up forever, with "discovery" requests, endless depositions, and baseless counterclaims. If we so much as threatened to sue, it would rain lawyers on our office like the plagues of Egypt.

Although no one in the industry would speak out publicly about this disgrace, the press was another matter. They could smell blood, and blood sells newspapers. The first to break the story was Andy Pollack, a technology reporter for the *New York Times*. Just two days after the demos, he ran a piece in the business section calling into question the legality of Microsoft's competitive practices. GO was mentioned in the lead. I was quoted as saying that people should think twice before showing their confidential ideas to Microsoft, but was careful to say that I did not know specifically whether Microsoft broke any laws. Soon after this piece ran, I got a call from Gary Kaye of ABC News. He told me that the FTC had just gotten permission to widen their probe of Microsoft, and wanted to know if I was prepared to comment on the subject on camera. I declined.

But after hanging up the phone, I realized that although I wasn't inclined to go after Microsoft in court, getting the FTC to do it for me was another matter entirely. The problem was that I couldn't just call up someone at a federal agency and tell him to get on the case.

Fortunately, I didn't have to.

Exactly four weeks after the demos at Esther Dyson's conference, John Croll, our in-house attorney, bounded into my office. A short man with bright eyes and a blooming bald spot, he had the temperament of a bantamweight prizefighter. He was winded from running down the hall.

"We got the call!" he shouted.

"What call?"

"The FTC. They told me they're looking into allegations that Microsoft is using its dominant position in operating systems to take over the applications business."

"John, sit down," I said. "You're hyperventilating." But he was too excited to pay attention to my picayune concerns. Corporate lawyers spend most of their time on boring matters like letters of intent, licenses, and contracts, and this was a chance to do something exciting.

He reluctantly took a seat, then popped up again. "Anyway, they want us to fly to Washington and testify next week — in secret."

From then on, John spent virtually all his waking hours preparing for the trip. He collected every shred of documentation that might bear on the matter — e-mail messages, engineers' notebooks, correspondence — and organized them into a carefully indexed chronological account of our dealings with Microsoft.

Two days before our trip, I was sitting in a meeting with a senior executive of one of the smaller Korean computer manufacturers. This was to be a routine dog-and-pony, where we would give him a demo and pitch him on building a pen computer that could run Penpoint. But I learned that this company was different from the other hardware manufacturers who were parading through our offices two or three times a week: it was not yet in the business of building DOS-based PCs. When the conversation turned to Microsoft, the executive grew more animated.

"I don't want to do business with them," he said matter-of-factly.

I was surprised, to say the least. "Why is that?"

"I don't like their pricing arrangements. To get the best price on their operating systems — one that would allow me to compete with other manufacturers — they want me to pay a royalty on every Intel-compatible CPU that I ship, whether or not I put their software on it."

He looked so angry, I wasn't sure I understood him correctly. "You mean if you license Penpoint from us and put it on a computer that contains an Intel chip, you would have to pay Microsoft a license fee as well as paying us, even though you were only selling our software?"

"That's correct. So I don't want to do it. That's why I'm here, to see if you have an alternative I can use."

"But that's outrageous. It means that the customer has to pay twice for our software." Microsoft had such strong control over their operating-system customers that it had, in effect, found a way to levy a 100 percent tax on Penpoint, which went right into their pocket. "Surely, this is illegal," I said.

"It's illegal if you tie the sale of one of your products to another of your products," he said. "But Microsoft is tying its products to the sale of *Intel's* products."

I was still doubtful. "Look, we've pitched Penpoint to lots of manufacturers, and no one has mentioned this to us."

"That's because they probably already have such a license with Microsoft, and the deal requires them to keep the pricing structure confidential."

Suddenly everything fell into place. In reasonable volumes, both Pen Windows and Penpoint should cost manufacturers about $50. But I had recently seen one of our licensees' price list for its upcoming pen computer product, and the company had inexplicably priced Penpoint at almost twice as much as Pen Windows. We assumed that this was some sort of mistake, and our marketing staff was in the process of trying to straighten it out. This wasn't a problem for IBM, because it already had rights to most of Microsoft's operating-system products. Other manufacturers had expressed a high degree of interest in Penpoint, but the plain fact was that other than IBM, NCR, and Grid, none had signed a license with us yet.

In an incredible coincidence, my assistant stuck her head in the room as I was pondering the impact of this revelation. "Sorry to interrupt, but there's a call I think you might want to take. It's Bill Gates."

I thanked the visiting executive for coming and excused myself from the meeting. On the way back to my office to take the call, I collected John Croll so he could listen in. "Be careful what you say," John said. "It can be used against you later, in court."

Gates sounded distraught. "I just want you to know that I've seen your comments in the papers. If you think we've done anything wrong, I'd be more than happy to provide whatever information you want on Pen Windows. I don't want to see this thing played out in the press."

I was noncommittal. "Bill, I understand your concern about comments in the press, but you know what it's like. They love to build things up."

"I'd be happy to personally come down and go over this with you if you think there is any problem at all."

"I appreciate the offer, but let me think about it. What I'll do is instruct the staff not to comment to reporters. We're simply considering our options. We aren't going to go off half cocked and file a lawsuit without looking at the matter very seriously."

"I just hope you understand what you could be getting into. You know our 'look and feel' dispute with Apple?" Gates was referring to one of the longest-running intellectual-property battles in the history of the industry. Back when Apple had first developed the Macintosh, with its graphical user interface, Microsoft was one of the earliest companies to agree to build badly needed applications. But before the applications were released, it was rumored that Gates had pressured John Sculley into granting him a license to adopt the Macintosh's graphical look by threatening to withhold the products. Sculley gave in, thinking there was little that Gates could do with the rights. Years later — after Microsoft introduced Windows, which copied many elements of the Macintosh — Sculley realized the seriousness of the threat resulting from this decision and filed suit, claiming that the license covered only an earlier version of Windows. Apple ultimately lost the suit. Now Gates spoke with restrained emphasis. "I've spent over four million dollars so far on that one lawsuit!"

He had couched his argument as friendly advice, but we both knew what he was talking about. There was no way a small startup company like GO could possibly afford the cost of legal action against Microsoft.

"That much, huh?" I wasn't going to get into an argument with him, so I changed the subject. "Maybe the first step is for you to allow us to look over a copy of Pen Windows in some detail, if you're willing."

"I'll have one sent down right away."

After we hung up, John narrowed his eyes suspiciously. "He probably got wind that we were about to visit the FTC."

I said, "He did seem to be on his best behavior."

Over the following weeks, Gates and I exchanged several letters stating our respective views of the interactions between Microsoft and GO. Gates's first letter emphasized that Microsoft had long been interested in handwriting-recognition software using the pen as the primary interface. He said that during his visit to GO's office in 1988 we had discussed the fact that Microsoft was working with several companies on systems software for handwriting machines, although

I don't recall such a discussion. Regarding my concerns that Lloyd
Frink had been exposed to GO's confidential information and was
subsequently assigned to the Pen Windows project, Gates stated,
"GO knew that we were seriously considering putting handwriting
recognition into Windows ... You were careful about what you
showed us and told us ... Lloyd Frink saw no code at all [and]
wasn't permitted to take any documentation, any software, or the
early version [of the Penpoint software development kit] out of your
offices."

When I wrote back to disagree, Gates sent me some "corrobo-
rating materials" to support his recollection of the 1988 meeting,
including a copy of an e-mail message to his staff in which he
directed them to consider applications for Penpoint, "albeit pes-
simistically." He added, "Regarding Lloyd Frink's role, I can understand GO's confusion about his titular responsibilities ... Lloyd
believes he negotiated in good faith the merits of GO licensing
Windows for its operating system while entertaining GO's arguments why Microsoft should become a GO ISV and that he did not
misappropriate any trade secrets in the course of his discussions
with GO."

▣

As we crossed the Potomac River and the monuments of the capital
came into view, I remarked to John Croll how remote the personal
computer industry seemed to be from government influence. It was
as though we lived in the Wild West, where it was difficult for
authorities to enforce the law because of the distances and the lack
of a local presence. Today the problem is different, but the result is
the same: the government doesn't understand the territory, and the
technology moves faster than the authorities can act. Civilized societies work only because most people willingly obey the law, even
if they are unlikely to get caught when they break it. On the frontier
this social contract decays, leaving in its place a different rule: Do
anything you want, as long as you can get away with it.

On the way from the hotel to our appointment at 601 Pennsylvania Avenue, we passed a large, well-kept colonial-style house surrounded by a high fence and an expansive lawn. I thought it a bit
odd that an old mansion like this could survive the onslaught of

government buildings and urban sprawl. "That's nice — what is it?" I asked the cabby.

He looked at me skeptically in the rear-view mirror before answering. "That's the back of the White House."

Our meeting took place in a marble and granite building that had recently been completely renovated on the inside. We were met by a young black man named Norris Washington, who was in every way the epitome of an earnest public servant. Despite the spare, modern decor, there were file boxes stacked everywhere, narrowing the generous corridors to a single lane.

"Are you just moving in?" I asked over his shoulder.

"No. They just didn't allocate us sufficient storage space, so we have to keep our records in the hallways." He led us into a typical law-office conference room. In the center was a polished wood table ringed by shelves holding endless volumes of leather-bound books with inscrutable titles. A staff economist and a technical computer consultant were waiting there for us to arrive.

Washington began the meeting by stating their objectives and the permitted scope of their investigation, as though reading from a book of rules. "We're ready to listen to anything you want to say. I want you to know that everything you tell us will be held in complete confidence, both from the press and your competitors. However, you are free to disclose our conversation to whomever you wish. Our ultimate goal is not to find out whether Microsoft is damaging a specific competitor, but whether there is a pattern of abuse indicating that they are damaging competition. If this is the case, it means that the law has broken down, and that's when the Justice Department is supposed to step in."

I began with a demo of Penpoint. It soon became obvious from their questions that they knew a lot about operating systems, the personal computer market, even about GO in particular. "We read the trade press religiously," Washington explained. Following the demo, John Croll began a chronological review of our dealings with Microsoft. The more detail he provided, the more interested they became. "Let me understand the sequence here," Washington would say. "This e-mail message was sent two days after the first meeting . . ." We were barely through half of John's review by one o'clock.

John and I had skipped breakfast, and were starved. "Would you folks like to take a break for lunch?" I asked. I had assumed that as government workers, they would keep strict nine-to-five hours with a one-hour lunch break, but they wouldn't hear of it. They waited for us while we bought some candy bars at a nearby vending machine.

As the afternoon wore on, I began to worry about missing our flight back. I had no idea they would be so intensely interested in our story, but it was clear that they had pretty much drained us of useful information. "We have to finish up, but if it will help your investigation, we'd be happy to return at another time," I said by way of conclusion. "But you're welcome to keep a copy of the materials we brought." Washington was delighted.

The three of them walked us to the door. While I waited for John outside the men's room, the staff economist approached me, and he let down his guard for a moment. Apparently, reviewing our tale was more than he could bear in silence. "It's incredible that no one has been able to stop Microsoft. This looks like a textbook case of abuse of monopoly power." He looked around to see if anyone was listening. "The problem is going to be convincing the FTC commissioners to take action."

About a week later, Norris Washington called me to thank us for helping them. "When we get to the next stage, we may need to issue a formal subpoena for you to produce these documents."

"No problem, we're happy to be of service."

"But right now, I don't know if we can use what you have."

After that long meeting, and their rapt attention, I couldn't believe it. "Why not?"

"At the moment, we're looking into the question of whether Microsoft has used their domination of the operating-systems business to take over the applications business. But in your case, they used their position in the applications business to their advantage in the operating-systems business."

I was stumped. He was quite correct — there were so few people in the operating-systems business, it would be virtually impossible to prove the "pattern of abuse" Washington required. To convince the commissioners, he had to stick to situations in which he could present lots of examples.

"If we widen the scope, I'll certainly call you again right away," he said, ending the conversation.

I broke the news to John. "Looks like Big Brother isn't going to ride to our rescue. We're on our own."

"There's never a cop around when you need one," he said.

▣

"Sometimes I think the loftier the title, the more foolhardy the executives get," said Carol Broadbent. She was talking about Charles Exley, NCR's chairman of the board, but she turned a little red as she said this, remembering that I had just been kicked upstairs to the job of chairman of GO when Bill Campbell joined as CEO. "Exley is apparently insisting that they announce the NCR 3125 on June twenty-fourth, three full months before his product team recommended. And he approved big bucks on this one. They're going to do a major production at the Hudson Theater, off Times Square. That's the same week as PC Expo." The 3125 was the name NCR had selected for its pen computer. By all accounts, it was a dynamite machine — a powerful 386-based design packed into an attractive three-pound package, far lighter than any other product on the horizon — that could run either Penpoint or Pen Windows.

I felt obliged to explain. "He's probably up to something and wants to position the company as forward-looking." We didn't know it at the time, but it seems likely that the NCR management team was running scared. AT&T was making a hostile bid for NCR, which put their jobs in jeopardy. "Don't fight them on this," I said. "Just find out what we can do to look better than Microsoft, and then let's make it happen."

This was easier said than done. Since Microsoft was one of its most critical suppliers, NCR deferred to its wishes at every turn, giving us short shrift. We had taken pains to train NCR's newly recruited pen-computer sales force on Penpoint, and support them on sales calls. As a result, NCR was finding that its initial customers preferred Penpoint to Pen Windows by more than four to one. Unfortunately, this was the "wrong answer" as far as the executives were concerned. Like everyone else in the industry, they didn't want to disrupt relations with Microsoft. But salespeople like to sell what

the customers want, so there was considerable friction between headquarters in Dayton and sales reps in the field.

The first conflict came when it was time to plan the stage sets
for the NCR announcement. Penpoint was designed to work primarily in portrait orientation, which meant that the screen was
supposed to be taller than it was wide, like a painted portrait. Pen
Windows, however, was designed for landscape orientation, like a
normal computer screen — wider than it was tall. The staging plan
called for a giant model of the 3125 to drop down from the ceiling,
so that demonstrations could be projected directly onto it.

Marcia Mason, our marketing communications person assigned
to the project, dropped by my office. She was steamed. "I just talked
to the A/V company in New York that NCR hired to do the show.
It seems that Microsoft is insisting that the model be in landscape."

"You explained that Penpoint works in portrait?"

"Of course. They told me confidentially that their instructions
from NCR are that when Microsoft says jump, they're only supposed to ask how high."

Bill Campbell flew to Dayton and took the matter up with the
senior executives there. They agreed to build two models: one would
drop down horizontally to show Pen Windows, the other would
drop down vertically for Penpoint.

At important announcements, it is customary for participating
organizations to issue separate press releases supporting the main
event, usually containing a quote from a senior executive at each of
the companies. When the time came to draft ours, Marcia asked me
to call Alok Mohan, the vice president of NCR's workstation products division, for a quote.

"Alok, I'd suggest something like 'Most of NCR's customers are
selecting Penpoint for their mobile needs.'"

There was a long silence on the line. "I don't think I can do that."

"But we both know that's the truth. Why do you want to hide
it?" I wasn't going to let him off the hook that easily. "If the situation
were reversed, and it was Microsoft, you'd do it in a heartbeat."

"Look, I'm rooting for you guys, but I just don't think we should
extrapolate from a small initial sample," Mohan said. "How about,
'Many of NCR's customers are selecting Penpoint.'?"

I reluctantly agreed.

When the time came for us to review NCR's press release, it was another story entirely. Marcia and Carol pointed out that although Microsoft and GO were mentioned in the same sentence, the text was formatted so that Microsoft's name appeared on the first page and GO's fell on the second. "You're not seriously suggesting that this was intentional," I said. "Even Microsoft isn't *that* petty." They both looked at me silently with eyebrows raised.

I called NCR's publicist and suggested a few wording changes that would have the side effect of moving our name back onto the first page. When the next draft came back, the margins had been altered so that the same thing occurred again. Carol and Marcia thought it was hilarious.

"I still don't believe it's intentional." I called NCR back to see what I could find out. After some probing, the publicist said sheepishly, "I shouldn't be telling you this, but I was specifically instructed to format it that way."

Carol gave me her best "I told you so" look. As they walked down the hall, Carol shook Marcia's hand as though she had just won a bet.

▣

In many fields, there are behind-the-scenes outfits that will quietly provide desperate companies with competent assistance to meet an impossible deadline. The music industry has anonymous studio musicians who can fill out a recording session in a pinch. The Seventh Avenue garment trade relies on quick-turn shops just over the East River that can cut and sew an important order over a weekend. And the computer industry has special engineering houses that can design a circuit board, provide a prefab power supply, or debug low-level software drivers for disks, screens, and other peripherals on short notice. If you look quickly, you may see a cryptic message flashing when you first turn on your PC, with a copyright symbol and the name of a company other than the one that's on the front of your computer. That means your top-shelf brand-name computer wasn't all built under one roof. To save time and money, some of its designs were purchased from one of those faceless companies, rather than built from scratch.

The bread and butter of these back-door companies is what's called a clean-room BIOS. This refers not to the tidiness of their offices but to their guarantee that the engineers who wrote the basic input/output system for your IBM-compatible PC did not achieve this compatibility by copying or looking at IBM's own BIOS. Instead, these firms hired virginal engineers who had never opened up the BIOS book from IBM and shut them in a room where they could see only *what* an IBM PC did, not *how* it did it, in order to reproduce the behavior as closely as possible. This way, the manufacturers could not be accused of copyright infringement.

IBM single-handedly created this absurd industry in 1981 by publicly releasing all of its PC designs but retaining exclusive rights to the lowest-level copyrighted program, which became known as the BIOS. Thus, to build a PC that is compatible with IBM's, you have to create or buy a clean-room BIOS. The most successful of these BIOS companies are those that have gone the longest without being challenged in court. Foremost among these in the early 1990s was Phoenix Technologies.

It was clear from its first presentation at the PC Forum in Tucson that Microsoft had a big advantage over us with respect to the amount of additional engineering a hardware company needed to do in order to run Pen Windows. Since Penpoint was a completely new operating system, the manufacturer had to create, in essence, a completely new BIOS, rather than adapting one it already had. This head start was what allowed Microsoft to claim that twenty-one hardware vendors had already signed up to build Pen Windows machines.

To close this gap, I negotiated with Phoenix Technologies for several weeks following that event to make a special deal: GO would license to Phoenix the designs for the 386-based version of our pen computer hardware — which our engineering staff had been busy developing since completing the 286 version — and Phoenix would create an approved BIOS for this design, which it could offer as a package to get its customers into the Penpoint business quickly. Phoenix was delighted at the prospect of expanding its business beyond clones of IBM PCs. As soon as the ink was dry, we scheduled a joint announcement for June 4 at a hotel near the San Francisco airport.

As members of the press shuffled past silver coffee urns, I met nearby with the senior executives of Phoenix to discuss last-minute details of our presentations. So as not to make it too obvious that we were still preparing our remarks, I sat directly in front of them on the corner of a wrought-iron end table with a glass top. They seemed strangely uncomfortable, which I assumed was because they were unaccustomed to public attention for their craft. But that wasn't it at all.

"Jerry, I think you should know that we had to disclose our discussions to Microsoft," said the dignified CEO of Phoenix, Ron Fisher, who was South African.

I smelled a rat. "Why?"

"You know we work with them closely and have a number of licensing arrangements with them," Fisher said. "On their DOS documentation, for instance."

A big rat. "So?"

"Well, they asked if we would create a dual-boot BIOS that would accommodate both Pen Windows and Penpoint, not just one for Penpoint."

In principle, I didn't have a problem with this. Our agreement didn't prevent Phoenix from offering the same sort of product for Pen Windows, though it wasn't quite our original intent. "That's OK," I said. "But this isn't Microsoft's announcement. The purpose of today's event is to show that there's momentum behind Penpoint, right?"

Before Fisher could answer, there was an explosive crash, as though someone had thrown a rock through a plate-glass window. I instinctively looked around before realizing that the sound had come from underneath me: the glass top of the end table had shattered into razor-sharp shards, and I was helplessly falling into them.

Thinking quickly, Lance Hansche, Phoenix's senior vice president of corporate strategies, grabbed my arm and pulled me out of harm's way before I could sink in.

"You all right?" Fisher shouted, losing his usual British reserve.

"Yeah, I think so." I turned around, and it was obvious that I wasn't. A thick stripe of blood was rapidly growing down my pant leg. The entire Phoenix executive team escorted me to the men's room, where they bent me over the marble sink counter and pulled down my pants to inspect. I had a large gash, which pooled with blood as fast as it could be swabbed. Someone who looked like a tourist entered the bathroom, took in the scene, and immediately stepped out, perhaps assuming he had stumbled upon some bizarre San Francisco cult ritual. By this time, the audience in the hotel's ballroom was mostly seated, waiting for the show to begin.

"Do you think we should cancel it?" asked Hansche. At that moment, a uniformed guard arrived with a first-aid kit. I looked over its meager assortment of remedies. A plastic spool of white adhesive tape caught my eye.

"Here," I said to Hansche. "Take this and tape up the cut." This was easier for me to say than for him to do, since I didn't have to look at the wound. After glancing uncertainly at the others, Hansche did the deed, and I hobbled out to the line of chairs on the podium. Realizing that I couldn't sit down, the other speakers stood as well, to camouflage the accident.

I went first. Spurred on by pain, I gave an impassioned speech about the burgeoning demand for Penpoint, as evidenced by the need for Phoenix to offer a Penpoint-related product line. Then it was Ron Fisher's turn.

"As the leader in the systems-engineering business, I'm pleased to be here today to make an important announcement that will help accelerate our customers' entry into the pen-computing business. Phoenix has agreed with GO and Microsoft to develop a special dual-boot BIOS that will support either Penpoint or Pen Windows. This way, hardware manufacturers can get started on their projects right away, without taking a risk today on choosing one or another of these fine systems."

Microsoft had browbeaten them into watering down the announcement. Fisher was apparently about to tell me this when I fell through the table. I was getting pretty tired of it all. Surely there was somebody who wasn't afraid of Microsoft.

A few weeks later, at yet another conference where GO went nose-to-nose with Microsoft, I ran into the director of portable marketing for Phoenix. He looked dazed, as though he had just been mugged.

"I think I need a psychiatrist," he said to me.

"My rates are reasonable, have a seat." I gestured to a folding chair in a corner of our exhibit booth.

"I just talked to Microsoft. They've decided to withdraw support for our dual-boot pen BIOS." He stared down at his shoes. "All that hard work to set this up, announce it, and brief our customers — and they pull the rug out from under us without warning."

After what Phoenix had done, I wasn't very sympathetic. "You've got to watch it when you work with them," I said. "Why do you think they did it?"

"I guess they figure that backing out now will hurt you more than it hurts them."

Since the dueling demos at the PC Forum, I had faced off about every two weeks, at one conference or another, with a speaker from Microsoft. Each time, we played the same cat-and-mouse game. I would try to distance Penpoint from Pen Windows — "maximize our product differentiation," as the marketing folks would say — and Microsoft would try to cozy up as close as possible, knowing that the more similar we appeared, the more likely that customers would choose its product. GO had long since abandoned the slogan "The Pen Is the Point" — which was intended to educate people about the overall potential of pen computing — and developed more focused, Penpoint-specific messages. At each event, I would emphasize a new element of difference. Between events, Microsoft would work to modify its demos and slides to copy GO's points in time for the next confrontation.

The New York air was uncharacteristically cool and sweet on Sunday evening, June 23, the night before the NCR announcement. A considerate breeze had blown the litter on West Forty-fourth Street into a neat ridge flanking one of the curbs, in preparation for its Monday morning street sweeping. The Hudson Theater, lovingly restored during a renovation of the neighboring Macklowe Hotel, occupied a prime Broadway location, but had a seating capacity more suited to an off-Broadway play. As a result, in 1991 it was mainly used for one-shot affairs such as concerts and private corporate events.

The production company hired for the NCR 3125 announcement was all too happy to have a virtually limitless budget. The NCR marketing department had given the producers a number of scenar-

ios to illustrate the use of pen computers — in an office, an auto
repair shop, a hospital. The production company designed a magnificent multilevel stage set as a backdrop for their live, dramatic
interpretations, in the style of a Broadway show. The centerpiece
was a sleek lectern mounted on Plexiglas legs so that it appeared to
hover in midair, flanked by twin glass TelePrompTer screens.

Because of the tight schedule and the complicated staging, all
visuals and speeches were carefully prepared and synchronized in
advance. The director wasn't about to have the show loused up by
some computer executive fussing with his slides or changing his talk
at the last minute. To assure a smooth performance, all demos had
been videotaped, edited to look their best, and transferred in sequence onto a laserdisk, from which they would be projected onto
various movable screens during the show. Since we were scheduled
to do a dress rehearsal that evening, at least I knew there wouldn't
be any surprises this time.

"Jesus, this looks like the Academy Awards," Marcia Mason said
as we entered the theater.

"It's more like Jerry Falwell preaching about the Second Coming," I said. "The 3125 won't be available for another six to nine
months at least." In fact, there were only about half a dozen working
3125 prototypes in existence, and none of them were in the theater.

As I waited for the rehearsal to begin, the actors practiced their
lines, attempting to portray people engaged in activities that they
didn't fully comprehend. Once the director's assistant had rounded
up all the participants, we ran through the show one segment at a time.

When my turn came to rehearse, Jeff Raikes, once again the
designated Microsoft representative, leaned nonchalantly against a
pillar in the dimly lit theater with his arms folded, paying close
attention. For this event, I had selected three words as my theme:
"Simple, Familiar, Consistent." I would illustrate each of these in
Penpoint, as a subtle contrast to the vagaries of Pen Windows. "Graft-
ing a pen onto an existing system makes it more complex, not
simpler," I would say.

Then Raikes and I switched places. "Compelling and Compatible" was his theme. Apparently, Microsoft had decided to adopt this
phrase in an attempt to turn around a stinging rebuke that had

appeared in the press following the PC Forum presentation: Penpoint was "compelling but not compatible," while Pen Windows was "compatible but not compelling." Raikes's meaning was plain: you can get it all from us; don't bother with Penpoint.

Next up was Vern Raburn, the CEO of Slate Corporation. On his way to the podium, he paused to give a warm hello to Jeff Raikes. Marcia whispered to me, "Did you know they were friends? I overheard Vern talking about playing golf with Raikes."

Since Raburn was also on the conference circuit, I already knew that I could no longer count on him to take a strong pro-Penpoint stand, as he could plainly see that Microsoft was creating a credible alternative. At each event, he had gradually changed his position from staunch GO support to a more neutral tone. As far as anyone at GO knew, his entire development team was working on only Penpoint applications, yet he was publicly positioning his company to become the Switzerland of pen computing. "You can do pen-centric applications on any operating system," he had said on a recent panel in Atlanta, to my dismay. But his real strategy was about to emerge.

First he ran through a few examples of PenApps, the Penpoint application he'd built out of the forms software I turned over to him just twelve months earlier. Then, to my horror, he showed a version of the same product running under Pen Windows. "Now customers' investments in development today are protected no matter which operating environment they ultimately choose to implement," Raburn said. He was making a play to put a common layer over both systems and thereby take control of both Microsoft's and GO's customers. Raikes was beaming. He knew that Raburn was bulldozing over the best parts of Penpoint and the worst of Pen Windows, leaving a flat, undifferentiated plain behind.

As soon as he was done, I cornered him. "Vern, old buddy, if I remember correctly, our forms-software deal says that you can't release a product based on GO's original code on another operating system."

"Check the fine print," Vern said. "That restriction only lasted for ninety days after we released under Penpoint. Besides, it's not a product yet, it's only a demo."

By nine o'clock the next morning, a large crowd was packed into the sloping theater lobby. Despite the early hour, there was the distinct electricity of an opening night. NCR had persuaded several hundred attendees of PC Expo to come to New York a day early, promising them nothing less than a chance to witness the birth of the next computing revolution.

The ushers assigned to the doors opened them simultaneously, and the crowd poured in. Each person wore a color-coded name tag: white for customers, red for press, blue for NCR personnel. VIPs had a small orange dot on their badges, and were escorted to a roped-off section in front. The audience settled down when the lights dimmed and a loud disco rhythm filled the air.

After bland introductory remarks by a few NCR executives, it was my turn to speak. As a counterpoint, I used the dramatic gestures and pacing of a TV evangelist, which played better in this theatrical setting. I returned to my front-row seat, and the announcer called Jeff Raikes to the podium.

Dressed in a flashy double-breasted suit, he stood casually at the podium and spoke with a slight smirk, giving him the faint aspect of a car salesman. When he got to the part of his speech about industry support, his eyes turned fiery. "Today we are excited to announce that over one hundred and twenty independent software vendors have committed to building exciting pen applications for Microsoft Windows." He paused as a daunting Who's Who of applications software scrolled by on a screen behind him.

I leaned over to Marcia. "Where the hell did this come from? It wasn't in the dress rehearsal. 'No changes' my ass."

Without a word, she raced off down the aisle to find the NCR person in charge of the show. I noticed that virtually all of our developers were listed on the slide, including Pensoft — which, I was confident, was building applications only for us. As Raikes returned to his seat, he glanced at me with his chin thrust out.

Marcia soon returned, and she was fuming. "She stonewalled me! Said that she was just following orders, there was nothing she could do. See if I trust her again!"

After the show, I called Michael Baum of Pensoft. If in fact his company had deserted us, I would be convinced that we were losing momentum.

"No way, Jerry," he said. "Pen Windows is seriously lame. It's mostly a demo."

"Then why the hell would your name be on the slide?"

Baum paused for a moment, then said, "The only thing I can imagine is that they listed everyone that showed up for their briefing in February or has requested a Pen Windows developer kit. I'll tell them to take my name off their list."

For some unknown reason, it was never removed.

10

·····························

The Spinout

IN THE SIX MONTHS since I had begun working with Bill Campbell, I learned by watching him what leadership was all about. The key skill is not in convincing people of your point of view with rational arguments, but, when circumstances require, in building a feeling of consensus in the face of uncertainty or adversity. Bill's strength was his ability to select a straight course through the swirling darkness, then create a deep emotional reserve in his team that drives them to victory, even when defeat seems inevitable. He metered out sufficient time for open discussion, then closed debate with a fatherly decision that all were expected to accept as their own, in the service of the greater good. His moral authority to work this miracle sprang from a simple fact: he was passionate about whatever he was doing, and he placed the welfare of his employees above all else, including his own.

Bill and I were like opposite poles of a magnet. I was cerebral, solitary, and analytical; he was gregarious, intuitive, and decisive. I did my best work in writing; he worked only face to face or on the phone. I was uncomfortable expressing my feelings around other people; Bill's trademark was to give people who wandered by his office a big bear hug. As our working relationship developed, we were joined together like yin and yang, each of us providing an essential component of the whole that could not properly exist without the other.

One of Bill's first acts as CEO was to establish a kind of corporate rhythm, a weekly sequence of meetings by which the company shared information and made decisions. This organizational pulse started on Monday morning at nine-thirty with "estaff" — a meeting of the executive staff, where large issues and strategic initiatives were discussed — and ended on Friday at four-thirty with the "comm" meeting, where the entire company gathered to communicate news, make announcements, and give demos and awards. Estaff started as "late" as nine-thirty because Bill, who usually arrived at the office around six, liked to spend the early morning talking to the East Coast, before people there went to lunch, and meeting privately with any employee who had a problem that couldn't be solved through normal channels. Following the comm meeting was a "beer bust," with drinks and munchies, a Silicon Valley tradition that Bill had followed since his Apple days.

For a man pushing fifty, Bill was in remarkable mental and physical shape. He worked out for an hour every morning before coming to the office while he read the trades, newspapers, and analysts' reports. Each evening when he was in town, he would plan a dinner with a business associate or visiting customer. But most of the time, he'd leave for the airport following estaff, often returning from Japan or Europe just in time for comm. He took great delight in his ability to outlast whomever he was traveling with, staying out late with customers and getting up early for pre-breakfast briefings. He thought it hilarious that I had to sack out at nine o'clock whenever we traveled to New York together. He'd head up to Columbia University for a night of carousing with old football buddies he had known since his days as the college team's coach. Wherever we went, Bill seemed to know everyone, and everyone considered him a personal friend. His family rarely saw him during the week.

After a harrowing several days of panels, presentations, and customer visits in the wake of the NCR announcement, I returned to San Francisco exhausted and discouraged. First thing Monday, I stopped by Bill's office. It was decorated like a cross between a Big Ten dorm room and a boys' clubhouse. Trophies and pictures were everywhere — mementos of sales club meetings, awards, snapshots of Bill and his freckle-faced son with sports celebrities. There were Nerf basketballs and a pyramid of baseball caps embroidered with the names of big league teams.

By contrast, my office was spare and sterile. A fraying gray futon served as a couch, a reminder of the company's humble beginnings south of Market Street. The main decoration was a picture frame I had purchased on a lark at a nearby drugstore. It came with a snapshot of a generic wife and kid, put there for those unable to imagine what it would look like with their own family in the frame. I took perverse pleasure when people admired the picture, only to realize that I had never removed the promotional insert. For me, it had been a symbol of the loneliness of my life. Now that I was engaged, I crossed out the woman in the picture and added a caption: "One down, one to go."

Bill invited me in, and I flopped into a chair. "Bill, we have to talk about the competitive situation with Microsoft." He could see that I was demoralized. This brought out the best in him. He had gained a reputation in the company as "the coach," someone you could always rely upon for emotional support and practical advice.

He put his hand on my shoulder in consolation. "Tough time in New York, huh? Jerry, this is a group issue. Let's discuss it at estaff."

For the past several months, Bill had been weeding out the weaker members of my executive team and bringing in his own hand-picked replacements. First on his list of recruits was Randy Komisar, GO's chief financial officer and vice president of business operations. Randy was a walking human contradiction. On the one hand, he was a shaved-headed former rock promoter from Pittsford, New York, who biked to work on his BMW motorcycle dressed from head to toe in leather, except for brightly colored socks and Day-Glo sneakers that had pictures of naked women, ghosts, or whatever else might strike his fancy. Weekends, he would take off for the all-night club scene south of Market, to do God knows what amid a crowd of transvestites and bohemian performance artists. On the other hand, he was a graduate of Harvard Law School who had worked for a blueblood Boston law firm. An accomplished, hard-nosed negotiator, manager, and administrator, Randy had all the skills of a high-powered lawyer and CPA rolled into one. He was a scrupulously honest and self-aware individual and could converse knowledgeably about psychedelic drugs or accelerated depreciation schedules, depending on the person he was with. Working with him was pure pleasure.

Another of Bill's recruits was Mike Homer, formerly director of marketing at Apple USA and Sculley's personal technical consultant.

A brilliant tactician with an encyclopedic grasp of all aspects of computer technology, Mike had a round face and a high-pitched voice. His lumpish body was thoroughly discordant with his razor-sharp mind, as if a wicked witch had condemned a brilliant scientist to living out his time trapped in the body of a cherub.

When I entered the main conference room for the estaff meeting, Mike, Robert Carr, and Kevin Doren were discussing the various methods of shooting paper clips by rubber band into the waste-basket. Randy was fine-tuning the position of his chair so that he could safely lean his head against the windowsill while putting his feet up on the table.

"Hey, did you hear the news?" Randy barked as I sat down. "Microsoft just took a lease on the entire ninth floor of the building — four floors below us!"

"What the hell are they going to do there?" I asked.

"Probably set up a listening station," Robert said. He was only half joking.

"I found out when I rode up in the elevator with Bill Gates for the office dedication," Randy said.

"This is great, this is really great," I said sarcastically.

Bill called the meeting to order. "Jerry, why don't you start us out."

I gave a brief report of the NCR announcement and my other experiences in New York. "Look, guys, this situation is ridiculous. We aren't on a level playing field. We aren't even on a tilted one. First Microsoft copies our stuff. Then they charge a 'tax' on sales of our products. And then they call the shots in forums where we should, at the very least, be on an equal footing. Our customers are afraid to speak out in our support. What the hell should we do?"

Bill looked around the table for ideas. Seeing that there weren't any, he said, "It's simple. We should go where they aren't."

I was exasperated. "Like where? They're everywhere! I check for Bill Gates under my bed in the morning."

Just then, Kevin pointed out that two window washers had just hoisted their platform into view outside the conference room window. "I suppose they work for Microsoft too," he said in jest. None-theless, Randy got up and closed the blinds.

Mike Homer had been thinking quietly. Then he said, "They're tied to the Intel architecture, we aren't."

"You're seriously suggesting we port the whole goddamn system to another processor?" I asked.

Robert lit up. "Why not? Look, Microsoft has to deal with all the baggage of DOS and the installed base. We've been talking for years about how portable Penpoint is. Let's pick one of the RISC processors and build a machine they can't touch."

It was a bold move, fraught with peril, but it just might work. "Intel will have a cow," I said.

Randy was ready with an answer. "Our contract with them says we have to offer Penpoint on Intel, but they can't force us to support it if no one wants it."

"They're unlikely to come after us in any case," Bill said. Then, "Anyone have a better idea?" Everyone was silent. "OK, let's spend the week looking into this, and finalize it before the comm meeting on Friday. Mike, you figure out which chips are realistic options. I'll get the KP guys to saddle up their horses. Randy, you check our contracts with Intel and IBM for pitfalls. Robert, get your managers to size up the project. Jerry, you quietly test the waters with some of our most trusted ISVs."

The news raced through the company as Robert fired up his team to work on a plan. It was as though Bill had sounded a wake-up alarm that rallied everyone from the receptionist to our Japanese product testers. By Friday, it was a foregone conclusion that somehow we were going to do it. When Bill announced the strategy to the entire company, there was enthusiastic cheering and applause.

At the beer bust I found myself face to face with Robert and Kevin. "Jesus, almost four years into this and we're still trying to find a workable strategy," I said. "I can't believe we're going to roll the dice again."

"Hey, as long as someone's willing to stake us, what the hell," Kevin said.

"Are you really prepared for another year or more of balls-to-the-wall development?" I asked Robert.

He paused thoughtfully. "Not really." Then he chugged down the last of his beer in several loud gulps. "OK, now I am." The three of us laughed nervously.

"John, why don't you rest a minute while we review what you've got on the board already?" Bill Campbell had invited the outside board members over to discuss the RISC issue with himself, Randy Komisar, and me. John Doerr had a habit of popping up at board meetings and writing lists of potential global partners on the whiteboard. On this occasion, his ideas for alliances were so broad that they resembled a tutorial on World War II.

Since the company had just about outgrown the entire fourteenth floor — with 130 people on the payroll — we were once again out of space. To commandeer a small corner office that had recently been converted to a makeshift conference room, Bill had to barter with a group of documentation editors who were eager to formulate their final recommendation on the optimal size of margins on a page. Bill offered to pay for their dinner if they would delay their meeting until six o'clock. Still fresh from lunch, they reluctantly accepted the bribe.

Only Vinod Khosla and John Doerr could make the meeting. David Liddle was in the middle of a behind-the-scenes battle to save his company, Metaphor, from being sold out as part of a deal IBM's Jim Cannavino was making with John Sculley of Apple. But David had much more than his company at stake. At Cannavino's urging, he had put his personal reputation on the line with a very public project called Patriot Partners, backed by IBM's promise of $80 million in funding. Their ambitious goal was to place a universal "object-oriented" applications software development environment on top of all the major operating systems. This was an attempt to bring order to the chaotic operating-systems API wars and, not incidentally, to break Microsoft's stranglehold on the market. After learning of the deal, Sculley had offered Cannavino a chance to buy into a similar technology that Apple had been quietly developing for years, thereby defraying Apple's heavy expenses on the project. Cannavino was intrigued by the possibility of making a huge splash by forming an alliance between Apple and IBM, two former archenemies. As a consolation to David, IBM had offered to buy Metaphor outright, but only for about one half of what the company was previously worth. Faced with the likelihood that the new alliance would badly damage Metaphor's prospects, David had stoically accepted the inevitable, but was valiantly negotiating for a more reasonable price.

Ignoring Bill's request to slow down, John began drawing lines on the whiteboard between three columns of names, one representing European companies, one for the United States, and one for Asia. Since there was no eraser in sight, he had to use the heel of his hand, which by now was a mass of black, red, and green splotches from rubbing out and rewiring the connections. "We're going to need global partners to be successful," he said. Randy winced at me.

Vinod was concentrating on the board. "It looks to me as though there are two stable camps," he said coolly, hands folded just beneath his chin. "And they hinge around two basic RISC-chip technologies, the ARM chip and the Hobbit chip. The ARM chip has the broadest reach. We could join forces with Active Book in England — who is just about ready to release their ARM-based pen computer in Europe and has approached KP for money — and Matsushita in Japan, who I believe has taken a license to manufacture the chips. Since Larry Tesler is also using ARM for Apple's Newton project, there's a lot of momentum behind the chip now. The Hobbit is more of a stretch. Right now it's an orphan, but with AT&T behind it, we could easily get their usual partners — Olivetti in Europe and NEC in Japan — behind it pretty quick."

Randy hadn't been around long enough to be comfortable with John's aggressive style and Vinod's bold strategies. He leaned back and folded his arms behind his head, as a counterpoint to John's and Vinod's growing intensity. "I'm all for international cooperation, gentlemen," Randy said, "but we're bleeding cash, and bad. So let's let the air out of our shoes. At our current rate, we'll be broke in late August. The Hobbit may be an orphan, but AT&T is one hell of a rich uncle." He put his feet up on the table, revealing on the bottom of his sneakers a psychotic array of wavy ridges with the words "Air Mowabb" in the center.

"You're one to talk," John said, pointing to Randy's soles.

"Check out his socks," Bill said, then returned to the topic. "IBM is beating us senseless to stop competing with them in the hardware business. Maybe what we can do is spin out our engineering group — maybe even sell it to AT&T — and agree to port Penpoint to the Hobbit. This way, we can save the jobs of our hardware team and get the new financing we need."

"Don't forget," I added, "that Tesler ditched the Hobbit chip in favor of the ARM, leaving AT&T high and dry. We could be just the ticket to save AT&T's project."

Vinod wasn't going to give up on the ARM so easily. "But if we convince AT&T to switch to the ARM chip, we could hit a home run. We could get Active Book, Matsushita, *and* AT&T."

John rubbed the heel of his hand with a handkerchief. "Bob Kavner, a group executive at AT&T, has a keen interest in communication devices. He's worked closely with Bernie Lacroute, who has just joined Kleiner Perkins from Sun. Let's ask Bernie if he would be interested in starting such a venture with Kavner and AT&T based on the ARM chip."

It turned out that KP had been in discussions with Active Book for some time, just as we had been quietly sounding out AT&T on developing some sort of partnership. Nonetheless, Bill, Randy, and I were stunned into silence. It seemed implausible that AT&T would ditch tens of millions of investment dollars at the urging of one of the KP partners. Yet John seemed quite serious, and Vinod was nodding his head in approval.

John was eager to charge ahead. "I'll get the ball rolling." He bolted down the hall to find a phone, presumably to call Lacroute, with Vinod close behind.

Bill looked bemused by this new turn. "I guess the meeting's over. Let's let them give it a try, then we'll get this thing back on track."

Randy shot me a skeptical look. I shrugged and said, "Look, they've made more outlandish suggestions than this — and been right."

Soon Lacroute was on a plane to AT&T headquarters in New Jersey, to put this proposition to Bob Kavner. It was no surprise when he returned empty-handed. John Doerr called to give us the news.

"Kavner wouldn't go for it," he said over the speaker phone in Bill's office. Randy mouthed the words "Of course," then adjusted his chair so he could balance his head against the wall. "So I called Active Book," John said, "and asked if they would switch to the Hobbit instead, if we agree to fund them."

Randy grabbed a corner of the desk as though someone had tried to tip his chair over. This was equally unlikely. Active Book was just about to release its own ARM-based product. Not to mention that the CEO, Hermann Hauser, was part of the team that had developed the ARM in the first place.

"And?" Bill said, as straight as he could. I was clutching my stomach, trying not to laugh out loud.

John paused, then said, "And they agreed. So Bernie's now working with Kavner, who's interested in funding the whole GO hardware engineering team, combining it with Active Book, and starting a global GO spinoff — if my partner Bernie will run it."

Randy snapped his chair upright with a bang. I was so shocked that I stood straight up. None of us could speak.

"Hello?" John said. "Are you guys still there?"

Bill moved his jaw up and down several times, trying to get his mouth into gear. "Uh, yes, we're here, John." He couldn't think of anything else to say.

John continued, making the matter sound entirely routine. "We need to get a proposal in front of them ASAP. Randy, can you work something up?"

"We can do that," Randy said, regaining his composure. "You'll have it first thing Monday."

Bill hung up the phone. Randy shook his head and said, "Gee, Toto, I don't think we're in Kansas anymore. Bill, if I ever question these KP guys again, kick me."

Bill faked a practice swing with his foot.

I walked over to the window and stared out at the Pacific fog rolling in over the western hills. "I just don't get it. Why on earth would Hauser tear up his product plans and go for such a wacky deal?"

"There's only one reason on God's green earth," Bill said. "The poor guy must be even more desperate for cash than we are."

One of the lucky breaks I had at GO was in hiring an executive assistant who cared deeply about people. Holli Maxwell would listen sympathetically to the woes of employees, investors, customers, and ISVs, then take on their problems as if they were her own. More than she could imagine, her actions helped set a tone of service and respect throughout the organization in a way that no written policy or corporate mission statement could.

One day, I walked by her desk to find her sitting quietly with tears streaming down her cheeks. "Holli, are you gonna be OK?"

Rather than answer, she handed me a letter, neatly written on lined paper torn out of a school notebook and addressed to me. She watched as I read it.

I hope this letter findes ya at a good moment. I wrote it many moths ago and am only now sending it to your company.

To Go computer . . .

I think it would be sad to just have big bussiness gain the power of this operating system. So I feel it would be a wise thing to seed your computer and operating systems to non business pepole, mostly artists and teachers, and seinor citizens, and disabled pepole. And althrough you might not choose to give me one, I would apprichiate you at least takeing this to heart and searching out these type of pepole to help, because these pepole have a lot to offer . . . may be not money but they are full of ideas and potentall . . .

I am servily learning disabled . . . My wife Lini who is a Kinder teacher and I run a play group, and I spend a lot of time designing things . . . my learning disability make it practicly imposabl to make sense or remember the cryptx and abstract think of D.O.S. . . . As it happens I believe to be compleatly taken over by your hole concept and product . . . so I think this is my dream come true . . .

School for me was, as might be amagend, painfull, if there was anything I could do for someone else esspicaly some one who is learning dissabled I would be happy. As I see it, technology for as much as it has done for bussiness it has greatly missed the opituneity to help a broard base of pepole . . . The Artist, teacher, learning disabled, sinor citizen, all these pepole have a very unique way of seeing and comunicating, and I think it is our responsibility as a society to provide them with the tools and then to listen to what they have to say . . . I could no way forcast great financial gains from this group but I could most certinly grantee there would be much to be learnt from them it they were to be incorparated into a longe range goal of commentment by you.

Joseph DeRuvo

Holli looked at me, wondering what I was going to do. I went directly into my office and wrote a response.

Dear Mr. DeRuvo:

Thank you for your thoughtful letter. I can see that your disability does not prevent you from thinking clearly.

While GO's technology will help large corporations work more efficiently, its main impact will ultimately be on individuals, as you observe . . .

In the modern world, access to technology is access to information, and information is power. This new freedom can benefit individuals as well as societies. Delivering computing power in a simpler, more accessible form can empower a broad group of people beyond those that use computers today. Our society tends to leave behind people who are not "plugged in." Perhaps our work can help a new group of people to be productive, to participate, to communicate, to belong . . .

In our rush to get a product to market, it's easy to lose sight of the deeper potential of our efforts. Our mission should be more than economic — money is not the only measure of our success. As Penpoint systems become available over the next few years, we will be sure that big business is not the only place they are tested and used.

Your letter was a great inspiration to all of us at GO. Be assured we will take your comments to heart.

Sincerely,
S. Jerrold Kaplan

Knowing Holli was concerned about confidentiality, I assured her it would be OK if she wanted to show the exchange of letters to some other people in the company before sending my response. When she gave me the final copy of my letter to sign, it had already been signed by every GO employee.

Making a business deal is like having sex: the more people are involved, the more difficult it is to consummate. Most negotiators prefer to make a single deal with one other party at a time — presumably a reflection of their personal preferences. That's why Randy was just the guy to pull together AT&T, Active Book, Kleiner Perkins, and GO at the same time, to form a pen computer hardware company based on AT&T's RISC chip, the Hobbit. Bernie

Lacroute had graciously committed himself to serving as CEO for at least the first year, as long as Kleiner Perkins got to invest, and he decided to name the new company EO — Latin for GO. Bob Kavner agreed to represent AT&T on EO's board of directors.

After weeks of intense negotiations, it looked as though the agreement was ready to be put to bed. Randy approached the deal with the firm but naive conviction that GO needed only an additional $10 million to get to profitability. This was based on an already outdated set of business assumptions and the questionable proposition that significant revenues would begin flowing from IBM and others by the fall. So he devised an arrangement whereby GO would get the needed $10 million as "advances on royalties," in return for a source-code license to Penpoint and cooperation in porting it to the Hobbit chip. Upon signing, $5 million was to come directly from AT&T, and the rest was due in installments from EO after GO met certain development milestones. When the deal was done, GO would have its money without giving up any equity; a new, committed hardware partner impervious to Microsoft's influence; and jobs for the people on the hardware engineering team. It all looked like a dream.

Believing that the agreement was complete, Bill left the office one balmy summer evening to go home and baby-sit for his newly adopted baby daughter. Randy was left to review the final terms on a conference call with the other three parties, who had gathered for this purpose at AT&T headquarters in Basking Ridge, New Jersey, the geographical midpoint of each of their offices. On the other end of the phone were Bernie Lacroute of KP, Hermann Hauser of Active Book, and Dave Atkinson, a tall, easygoing staffer of Bob Kavner's who had been assigned to handle AT&T's investments in new computer technologies. Normally, Atkinson's relaxed, affable manner made discussions with him feel like a Saturday afternoon at a Pebble Beach clubhouse. But in combination with Lacroute's tough negotiating style and Hauser's concern that he might somehow be cut out of the deal, Randy had his hands full.

An adroit negotiator himself, Randy knew that all good deals come right down to the wire. Otherwise, someone has left something on the table — and he liked a clean table. It was nine P.M. in San Francisco, and as eager as he was to escape and get some dinner, he knew that the others must have felt worse: it was midnight in New Jersey. Hauser planned to return to England the next morning, around the same time as Lacroute's flight in the other direction. They had resolved all the issues except one, a requirement that GO release future versions of Penpoint no later on Hobbit than on other platforms — that is, Intel's. Aware of our symmetrical agreement with Intel, Randy realized that this would dramatically complicate GO's product development, since both versions would have to be perfected and released in lock step. He was understandably reluctant to tie our hands.

"Sorry, but no way I can go with this," Randy said. "What I can offer, however, is that we will tell you if we can't meet this require-ment and give you the right to finish and release the Hobbit version yourselves."

Back in New Jersey, they were ready to explode, and Hauser lit the fuse. "What you're telling us is that you're not really serious about this, not really committed to Hobbit! You expect me to give up control of my company, throwing our lot in with you, but you can walk away if you feel like it?"

Lacroute echoed Hauser's point. "No version parity, no deal. This is it."

"Yeah," Atkinson agreed, "this is it. This is really it."

Despite the hardball tone, Randy concluded they were making a mistake. The right way to play this would have been for one of them to leave the room while the others loudly objected, so they'd have a graceful way out if they changed their position later. This was why Randy had let Bill go home.

He decided to stand firm. "Gentlemen, I guess there isn't much point in talking further. I'm going to dinner now. Thanks for all your efforts." To the amazement of the three men waiting to close the deal at the other end, he hung up the phone.

When I got to the office the next morning, I found Celeste Baranski and her husband, Mike Ouye — our directors of hardware and of software — barricaded in the office of our recently hired VP of operations, Paul Hammel. The three of them were to form the nucleus of the new ten-person EO engineering team, and had al-ready negotiated employment packages with Lacroute. (Kevin Doren had decided to stay with GO, as we still needed some in-house hardware expertise.) We were sorry to lose Ouye on the software side, but he and Celeste were a package deal. I was the only person they would allow in.

"What's going on, you guys?" I asked, squeezing into the room.

Hammel's eyes were red for lack of sleep. They stood out against his chalky face, giving him the drawn look of an apparition. "You've got to do something. Randy's being an asshole. He's blown the whole deal!"

"How's that?"

"Bernie called last night," Hammel said. "He told me it was off, because Randy wouldn't agree to maintain parity on the Hobbit. You've got to talk some sense into him." The three had whipped themselves up into a paranoiac state.

Celeste was on the verge of tears. "Jerry, you of all people know that we've been in this together from the beginning. Don't stand by now while Randy and Bill betray us."

I headed down the hall to see what I could do. Randy was relaxing in Bill's office, seemingly oblivious of the trauma he'd caused. "What's the scoop?" I asked. "Our people are pretty upset."

"Unfortunately, that's just Bernie's way of putting pressure on us," Randy said.

"Maybe he's just keeping his people informed about the negotiations." Since I wasn't the one on the spot, I could be more objective about the situation.

"Next, I expect he'll call Bill." Randy's face softened a bit. "I hate to see them so worried, but it will all work out fine. Do your best to calm them down. Celeste won't listen to me."

Sure enough, within hours Lacroute had called Bill and worked out a compromise along the lines of Randy's proposal, and the deal was done. But the hardware team never shook off the feeling that they were somehow cast aside, and this lingering suspicion was to taint GO's relationship with EO.

The Switch

"IF I'M GOING TO get shot, I want it between the eyes, not in the fuckin' back!"

I had never seen Bill Campbell so angry. He had just gotten off the phone with Kathy Vieth of IBM and stormed down to my office. He was so enraged his eyes were bloodshot. Bill could suffer just about any indignity with grace, except when someone had abused his trust. And Vieth had hit his hot button with a sledgehammer.

"Jesus, Bill, calm down." I couldn't imagine it was all that bad. After all, Kathy had been doing her best to support our cause.

"I just caught her in her car, leaving Apple."

"What are you talking about? I spoke to her on Friday — she was back in Minnesota for a week, visiting her brother." As a cost-cutting measure, everyone at IBM had been given the week of July Fourth off, a sort of forced vacation. Kathy had told me how nice it was to see her brother and how great the weather was back home. She had said that her beeper didn't work there, that it would be tough to get in touch with her, so not to try to contact her.

"Bullshit," said Bill. "I'm telling you I just talked to her. She's staying in San Jose!" By now Randy was in my office to see what the ruckus was.

Bill unloaded his tale from the beginning. "I get a call from one of my buddies at Apple, see, tipping me off that Cannavino and Vieth are in with Sculley. He's pitching them on dumping Penpoint and joining forces with Apple on Newton, along with a bunch of other joint deals they're making."

"Oh, Christ!" Randy's eyes narrowed. "Nah, I don't believe it. Not after all the promises they've made to us. IBM's difficult, but not devious."

"So I figure what the hell," Bill said. "I'll call Kathy's car. Sure enough, she was dumb enough to pick up the phone. You should have heard her — scared the shit out of her. I actually caught IBM red-handed, looking to sell us out."

"So what'd she say?" I asked.

"First she denied the whole thing. Then she fessed up about the meeting. I asked how it went. And she just said, 'I can't comment on that,' like I was some goddamn reporter. So I told her that if they were going to take a different path from us, that was fine with me, I couldn't dictate their business direction. But I want her to tell me to my face. So I told her she had till the end of July to make up her mind. Then I told her I wanted to talk to Cannavino about this, and she just about jumped down my throat. She said, 'Don't do that, I'll handle it.'"

Bill's conversation with Kathy Vieth must have shaken things up a bit. No doubt she contacted Jim Cannavino immediately. Within a few days, Cannavino had called Mike Quinlan out of retirement to deal with us, presumably thinking that we were more likely to trust him than Vieth, now that her credibility was damaged.

Quinlan came in to smooth things over and, ostensibly, to brief us on the upcoming joint IBM-Apple announcement. He was dressed for golf, and said he was on his way to a family reunion in Hawaii. He took the credit for patching things up. "I want you to know that I spent many hours with the IBM team evaluating Newton, and I convinced Cannavino to stick with Penpoint. I told him that by the time Newton could ship, Microsoft would most likely have taken over the pen market. Also, we've already promised Penpoint to a number of key customers. Besides, with all the other joint agreements — a new chip, a new object-oriented cross-platform operating system, a new standard for multimedia — we shouldn't try to boil the ocean in one shot." He kept using this unusual phrase over and over again. These other projects ultimately became known as the Power PC, Taligent's Pink, and Kaleida's Script-X. "So I want to reassure you that we're going to stick with GO."

To retain our attention, Quinlan revived a discussion that had more or less died after he left to retire: arranging a sizable loan to

GO as additional financing. "I can't make any promises, but confidentially, things are a lot looser than they were the last time around. It looks like we can earmark perhaps ten million." Randy looked as if he had just been offered money in return for a night with his wife. Nevertheless, he was destined to spend much of the next year in negotiations over the loan while IBM added ever more onerous terms.

After Quinlan left, Randy, Bill, and I regrouped in the conference room to compare notes. "It's so hard to believe they would consider dumping us — behind our backs too," Randy said.

For once Bill was the calmer one. "Actually, we're better off now that we know what kind of partner IBM really is. If this happened once, it will happen again. We've got to close the AT&T deal right away."

"You bet. I can't wait to tell IBM to get lost," Randy said.

Bill walked over and stood directly in front of Randy, leaned in dramatically, and put one arm on his shoulder. "It's not time yet to burn any bridges. We need to keep IBM on the hook for the time being." Randy sighed, nodding in agreement as Bill continued. "We've just got to suck their ass through a Flavor Straw until we don't need them anymore."

By October of 1991, the hoopla about pen computing had continued to escalate, despite the fact that there were no real sales in sight for anyone. After repeated delays, IBM and NCR had yet to release their pen computers for sale. The few prototypes they had in circulation were in constant use, on short-term loan to one company or another, creating an impression of overwhelming demand and shortage of product.

Once again, GO was running out of money, and the only real prospects were to agree to the IBM loan or to attempt another round of financing. Despite Mike Quinlan's optimism, IBM was as difficult to deal with as ever. Randy spent endless days in discussion with a newly assigned negotiator. Every time Randy thought they were close to a deal, the negotiator would change the ground rules. In its latest demand, IBM wanted the rights to use our notebook's "look and feel" on top of OS/2, DOS, Taligent, and Unix — for free. The negotiator argued that IBM was unlikely ever to use these rights, but the various development groups insisted that they be allowed to maintain a level of consistency across their product lines. IBM's strategy was simple: wait until we were out of money, and then we'd have to accede to its demands.

GO's alternative was to raise yet another round of financing from investors. We had just about tapped out the venture capital groups. After all, they had put up a total of $23 million, in addition to the $7 million from IBM and the $10 million from AT&T. To attract a new class of big investors, we had to spiff up our image and increase our public visibility. Months of going nose-to-nose with Microsoft and the announcements of several short-lived but highly touted competitors had reduced people's perception of us: we were seen as just one more independent contender in a crowded field. Fortunately, fall and COMDEX were just around the corner.

In the hopes of finally crushing us while they had the home-court advantage, Microsoft allocated a large COMDEX booth to Pen Windows: perhaps fifty by seventy-five feet of real estate in a prime location. As probably the single largest exhibitor, Microsoft had plenty of weight to throw around. Despite its strong recruiting efforts, I was pleased to see that it didn't have any more ISV applications to show than we did. And Microsoft's were plainly of lower quality, mostly minor modifications to existing Windows applications.

The heavy behind-the-scenes pressure continued. I asked representatives of one hardware manufacturer why they ran only Pen Windows on their machine. They told me privately that Microsoft had offered them special concessions and was willing to provide a senior executive to speak at their product launch, if they worked exclusively with Microsoft.

Unable to afford a real booth, GO took a small private room off the main hall. The rent was $53,000 for the week. To have the room vacuumed once a day cost an additional $1,000. To turn on the electrical plugs — which were already installed, but disabled — the fee was $150 per plug. The electricity itself was extra. It felt like the last days of the Weimar Republic.

What we lacked in funds our crew made up for with enthusiasm and imagination. To save money, they carpooled for the twelve-hour ride from San Francisco to Las Vegas, then worked tirelessly from dawn to dusk. They canvassed the aisles and halls wearing sport shirts with "Ask me about Penpoint!" printed on them, handing out passes to hourly shows in our modest room. The presentations were consistently packed.

It was against COMDEX rules to give demos outside officially rented spaces, so our guerrilla demo squad, armed with Penpoint computers, would lure prospects to the nearest hallway to avoid detection. To attract interested parties, the guerrillas would ostentatiously whip out their computers and pretend to take notes on some exhibitor's presentation. When they discovered a stubborn hardware manufacturer with a pen computer running only Pen Windows, they would borrow one of the manufacturer's units and run it back to our room, where a bearded, irascible engineer sat all day behind a curtain loading Penpoint, like a short-order cook in an off-brand burger palace.

My mission was to get our product mentioned in the press. It wasn't hard.

The first night, *PC/Computing* magazine held its Most Valuable Product awards. This lavish one-night Las Vegas show, which featured the magic of Penn and Teller, was broadcast on closed-circuit TV to the hotel rooms of the 130,000 COMDEX attendees. After being escorted to the show in a complimentary limo, I accepted one of the product awards on behalf of Robert Carr and the entire development team. Penpoint was praised at length in the magazine's next issue, which reached nearly one million readers.

The next morning, I did an interview with Gary Kaye of ABC News. Then, after spending several hours chasing down various reporters, I developed a new strategy. Borrowing a press tag, I set myself up in the press room, where everyone eventually came to me instead.

The following evening, I was asked to go to the *PC Magazine* awards — not to be confused with the *PC/Computing* awards, although both monthlies were owned by the same publishing company. This time I had to wait in line for a cab, a forty-five-minute proposition at best. Luckily, I ran into Steve Ballmer and Bill Gates, who invited me to hitch a ride in their limousine.

Ballmer looked a little embarrassed at this apparent extravagance. "The hotel provides it for free," he explained, "because we rent so many rooms from them." On the way to the show, they were all business. To my surprise, Ballmer was not reticent about briefing Gates in my presence about what he had learned from hanging around IBM's OS/2 booth. He listed, feature by feature and application by application, what worked and what didn't. "They've done a better job than I would have guessed at getting Windows apps running," he said. "But I still contend that full Windows compatibility is virtually hopeless in the long run." This had become IBM's approach: OS/2 would be a better Windows than Windows itself, since IBM had rights to the source code — for the time being.

Gates agreed. "It's only a matter of time until they can't keep up," he said ominously.

Ballmer turned his attention to me. "I imagine you're going through the same things we did with IBM."

"Probably." I kept a poker face.

"We were ultimately forced to make a hard decision," Ballmer said. "Whether to build the right product or do what they wanted. It was very emotional for us."

After a brief silence, Gates spoke up. "I hear that your system is quite a bit bigger than you thought."

I was amazed that he would know this well-kept secret. To our horror, the latest version of Penpoint was far larger than we had anticipated. In fact, it wouldn't even run in four megabytes of memory, so we had to load up the demo machines with eight. This would be a critical shortcoming if it couldn't be fixed, as it would drive up the cost of the hardware significantly. Most of our programmers were working night and day to figure out what went wrong, and to squeeze it back down. But virtually no one outside our shop knew of the problem, except for a few of our most important ISVs.

"Not particularly" was all I said.

The black-tie awards ceremony was even more glitzy than the one the night before. Lasers and confetti made it seem like a cross between a rock concert and a game show, and was clearly produced for the television audience. This time, Penpoint won the Standards and Systems Software Technical Excellence Award.

Later in the week, I attended an enormous IBM party celebrating the release of the latest version of OS/2. The only problem was that the product had been delayed by several months, and so wasn't being released at all. But that didn't stop IBM from staging a festival of self-congratulation, during which a stream of executives toasted the development team from the podium.

Afterward, I was cornered by a group of four or five Penpoint ISVs. They were plainly agitated. "Have you heard what IBM is telling us now?" one of them said. "They're saying Penpoint is for vertical markets only — like the State Farm application — and we should target our work toward their new secret project, Pen OS/2."

I said, "This is a joke, right?" But they all shook their heads in unison.

I immediately tracked down one of the senior engineers on the OS/2 team who had participated in a number of Penpoint evaluation meetings the previous year. "Rumor has it you're developing pen extensions to OS/2. Is this true?"

Fortunately, no one had bothered to tell him to deny it. "You mean our people haven't told you? That's why we're so adamant about getting rights to your 'look and feel' for our other operating systems."

I quickly found a bank of pay phones in the hotel lobby, and was on a conference call with Bill and Randy within a minute to relay the news. The greasy metal phone cord made a sound like a machine gun as I pulled it around the last plastic courtesy panel to get some privacy.

"Just stay cool," Bill said. "We still haven't closed on the loan of ten million. Then we can work on finding a real partner." I slammed down the phone with such force that I nearly broke the little plastic tongue out of the handset cradle.

As I walked away, I passed a hotel security guard lounging nearby. "What's the penalty for murder in Nevada?" I asked.

"Life — but figuring in good behavior, you're eligible for parole in ten years," the guard said. "It's better to hire someone. Want me to look into it?"

I laughed, because he seemed to take me seriously. It reminded me that this was only a game. A very big game, with serious consequences.

◰

Bill banged his fist on the table. "Financing has to be our number one priority. Let's talk about options. And don't mention layoffs on my watch — the next CEO can try that solution."

Robert, Kevin, and I knew the seriousness of our financial problems, and we had long since grown accustomed to living periodically on the edge of extinction. For the other executive staff, however, staring into the jaws of death was a new experience. They reacted like any sane person who suddenly faces his own mortality. Initially, they went into denial — continuing business as usual, planning off-site staff meetings, interviewing new hires, and visiting customers — figuring something would bail us out. Then came acceptance. Mike Homer had quietly started thinking about which of his people he could afford to lay off and which could easily find other jobs on their own. Eventually panic set in.

"Our options are pretty limited," Randy said. He was putting it mildly. We all knew that with almost two hundred people on the payroll, we were spending nearly $2 million a month, but it sure sounded different when you realized that AT&T's $10 million would last less than six months. With Penpoint Japanese, Penpoint Hobbit, and Penpoint Intel — not to mention applications development, ISV recruiting, customer sales, hardware manufacturer support, marketing communications, public relations, technical support, documentation, quality assurance, product marketing, field sales offices, GO Japan, training, human resources, legal, accounting, and facilities — everyone was working around the clock and still screaming for help.

Randy continued. "We can cave in to IBM on the loan terms, but this means handing them the right to rob us blind. We can ask our Penpoint licensees for advances, but they're likely to see it as a sure sign we're in trouble, and run the other way. Then there's AT&T. I suppose we could see if they're interested in making an additional investment."

Bill pursed his lips. "No way. Not while we're fighting tooth and nail to win the Telepen RFP." He was referring to a Request for Proposal issued by AT&T's general business systems division, to design an operating system and a standard user interface to run on a wide range of computer equipment planned for the future, including all pen-based devices.

Since the government had forced AT&T to divest itself of the regional Bell operating companies ten years earlier, the company's management had wisely granted substantial autonomy to the remaining divisions and fiefdoms in an effort to sort out what was making money and what wasn't. Not surprisingly, they found widespread duplication of functions and many competing projects. Now AT&T was paying the price, a virtual Babel of incompatible software and standards that operated the myriad switches, trunk lines, and phone systems sold and serviced by AT&T. When word of the EO investment had circulated throughout AT&T, related projects — some pure research and some more practical — surfaced like worms in a rainstorm. Determined to avoid chaos, the top AT&T managers agreed among themselves to establish one interface standard across all divisions.

The stakes were enormous in this winner-take-all game — it was nothing less than a chance to create the interface for the next generation of telephones. Bill Campbell had assigned GO's best people to working this opportunity. As far as we could tell, GO was bidding against Microsoft, which was offering a stripped-down version of Windows it was developing, and against an ultrasecret Apple spinoff called General Magic, which had a contract with AT&T's data transmission division to build a new generation of data communications software.

"The only other alternative is to try the financial community again," Randy said. "But it's likely to be tough during a recession."

"OK." Bill was ready for his call to action. "Let's get out there and scare up some new investors — somehow, somewhere." He paused and looked around the room. "Anyone have any last suggestions?"

"How about a bake sale?" Robert asked, biting into his muffin. "Let's see, at ten thousand dollars a cake, we only need to sell fifty a week."

No one laughed.

Armed with a fresh pile of awards and press clippings from COMDEX, Randy and I were ready to kick the financing into high gear. We put together a slide presentation for investors. "You know, this is pretty persuasive," I said. "We're technically the best. There's a lot of hardware products being developed that will run Penpoint. We're signing up licensees in Japan, IBM and AT&T are behind us, and the next version of Penpoint, the one for customers instead of just developers, is only months away." After working ourselves into feverish optimism, Randy and I hit the road like Hope and Crosby.

The first stop was Bloomington, Illinois. We gave a quick update to Norm Vincent and his claims-estimating development team. He was as cordial as ever. I apologized for not visiting for so long. "Norm, I hope you don't think we only come around when we need money."

"Jerry, don't worry. I understand what you're going through. I want you to know that we asked ADP to bid on building a Penpoint-based estimating system for us, but they came in and pitched us on working with Microsoft instead." Randy and I held our breath: ADP was a large systems supplier for the auto claims market. "Do you believe that? I sent them packing, of course. As you know, State Farm just doesn't like surprises."

He was as stalwart a supporter as ever. He took us to lunch with the investment folks. To our delight, they committed an additional $2 million, $4 million if we needed it. But this time they wanted someone new to set the price.

Randy returned to the office to see if there was any last hope of wrestling reasonable terms from IBM, while I crisscrossed Manhattan like a sailboat tacking upwind, visiting old investors and fishing for new prospects. Having long since outgrown the venture firms, I worked my way up the food chain — past the corporate investors and pension funds, finally calling on the investment bankers who usually took companies public. I soon found the hook to interest them: if they put some money in now, they would get the inside track on our IPO. Hungry for "product" (as promising young private companies are known on Wall Street), they seemed quite interested in landing GO as a client. They all read the same market research studies and didn't want to miss the boat in pen computing. Whereas most investors looked squeamish when they saw that we had already blown close to $40 million, the investment bankers seemed to take comfort from the larger than usual capitalization. As one wag put it, "You guys should be investment bankers — raising all that money while still retaining thirty percent for the common stockholders. Maybe it's time for us to lend a hand before you put us out of business." But no one was willing to set the price.

By the time I got back to the office, we had about $12 million in sight. It looked as though we might pull through — if we could find a lead and close up the financing fast enough.

Coincidentally, an old rugby teammate of Bill's happened to call, and he asked Bill if we needed any money. "I can put you in touch with a buddy of mine who manages a pension fund," he said. Within hours, I had a package off to a firm in Los Angeles whose partners had read numerous articles about pen computing and seemed grateful for an opportunity to invest. They soon committed $2 million, at $3.25 a share.

"What does that make the pre-money valuation?" asked Scott Sperling of Aneas, one of our earliest investors, when I called to give him the news.

"A hair over one hundred million," I said. We both knew this was an aggressive valuation.

There was a long pause, then he spoke hesitantly. "All right, we'll take our pro rata share if absolutely necessary to make this happen, but only if everyone else will. But I want you to know that this is the last time. From here on out, you're on your own."

"Scott, I understand. Thanks for your support on this one." I shivered as I hung up the phone.

⊡

Just after New Year's of 1992, we were ready to close on this round of financing. Final papers had been sent to all the investors for review, which were due back with their funds on Monday, January 6. But the Friday before, around six P.M., I got a call from our attorney's office.

"We have a little problem," said the paralegal assigned to the matter. "We can't issue a clean Opinion."

One of the reasons lawyers are so highly paid is that they are on the hook for making sure that everything is legal and proper. Their blessing — essentially a written statement that puts them right out front in case of an irregularity that leads to a lawsuit — is a letter sent to the investors called an Opinion of Counsel. It attests that as far as they can determine, everything is aboveboard. Since for the most part they are simply stating that they did their job, I had never heard of a problem in issuing an Opinion, unless there was blatant evidence of fraud or misrepresentation.

"Why the hell not?" I asked.

"Sandra Shmunis."

"Sandra who?"

"Shmunis. It appears that she may have been improperly issued options on ten thousand four hundred shares of GO common stock without a board resolution."

I was still stuck on the last point. "Who the hell is Sandra Shmunis?"

"I don't know," the paralegal said. "But you better find out quick. If this is true, all the closing papers that we drafted, sent to investors, and filed with the Department of Corporations have incorrect capitalization tables. And our state authorization for the closing expires next Wednesday, the eighth."

"What happens then?"

"We have to start again. New board resolutions, new closing papers, new filings . . ."

I knew what that meant. Another $20,000 in legal fees at least, and several more weeks during which the investors could change their minds. "Can't you go ahead and close, and we'll sort this out later?"

"Then the investors can sue you — and us."

"How about voiding the grant — if it ever happened — and reissuing it after the fact?"

"Then *she* can sue you — and us."

I started to panic. I searched the halls for someone who might have a lead on Sandra Shmunis. No one had any idea who she was. In desperation, I got up on a chair, looked out over a sea of cubicles, and yelled out, "*Has anyone ever heard of Sandra Shmunis?*"

A full minute went by, then someone wandered over and asked, "Isn't she the new person in tech support?" In a flash, I was outside the tech support room. It was locked. I called the head of the department at home.

"Sure, she started in October," the supervisor said. "But I believe she just left on vacation."

I coaxed him into finding Sandra Shmunis's home phone number, and I dialed it. I counted the rings. On the fourth, the answering machine clicked on. I was hosed.

Then I heard someone struggling with the receiver, which promptly issued a loud squeal because the answering machine was turned up too loud. When it stopped, I heard children crying in the background. "Hello, is this Sandra?"

"Yes?" said a timid voice.

"Thank God. This is Jerry Kaplan. You know, chairman of GO? Sandra, have you been issued your stock options yet?"

"I — I don't know."

I had inadvertently scared her. A recent immigrant from Russia, she wasn't prepared to get a call from the chairman questioning her about her stock.

"Did you get any papers from human resources?"

"Well, yes. Was I supposed to do something with them?"

She didn't realize she was supposed to sign the option grant form and turn it in. This would have triggered a procedure to add her name to a list of people whose options would be approved by resolution of the board at the next meeting.

I took a deep breath. "Sandra, would you mind signing the form and dropping it by the office tomorrow morning?"

"Of course. Is there a problem?"

"No, I'm just trying to be sure you get your stock the way you're supposed to."

She must have thought I was nuts.

The next morning, the lawyers sent a blizzard of faxes. We tracked down the board members over the weekend and had them sign a "Resolution by Unanimous Consent," thereby avoiding the need for a face-to-face meeting. Then new papers were transmitted to all the investors. By the morning of January 8, the last day we were legally allowed to sell the stock, we were ready to close. Or so I thought.

"There's one more problem," said the paralegal. "The new Japanese investors wired their funds two days ago, and due to the delay, they want to invest the interest earned in the company."

"How much?"

"It comes to a little over two hundred fifty dollars. We'd have to redraw all the papers, as this would increase the number of shares issued."

"Tell 'em no way. No, wait, I'll tell them. I know — issue them a gushing apology and wire them the amount in dispute."

"You want to issue an international wire transfer for two hundred fifty dollars?"

"You bet, if it will allow us to close. Just do it."

That afternoon, the attorneys deposited $12,935,503.70 in GO's bank account — after subtracting the two days' interest.

▣

While working on the financing, Kevin Doren accepted the job of securing licenses for certain key technologies that our hardware partners wanted us to use. One of these technologies, known as FFS2 (Flash File System 2), was a method for storing data in special memory chips that retain their contents when the computer's power is turned off. Intel and Microsoft had recently announced this technique, which they had developed jointly, stating that they would make it public in order to establish an industrywide standard.

Kevin started by contacting Intel, but learned that the technology was available only from Microsoft. After some effort, he succeeded in finding the person responsible for granting licenses to FFS2. Over the phone he was told that the written specifications would be made available to GO, but a license to the software that implemented it would cost $150,000.

"I thought it was supposed to be freely available," Kevin said.

"It is freely available — it's just not free," said the Microsoft representative. Kevin politely objected and hung up.

He was still fuming over the price when he got a surprise call from Lloyd Frink on the Pen Windows project. "You're negotiating with me now," Frink stated. "We might be willing to give you the FFS2 license in trade for the rights to some of your patents and copyrights." Kevin was astonished, not just that Frink would suggest this but that in a big corporation like Microsoft, his request for a simple file-format specification would so quickly become known to the group working on Pen Windows. "And one other thing," said Frink. "We also want a release from any claims arising out of my earlier visits to GO."

Kevin was temporarily speechless. Then he spoke. "Are these Microsoft's usual terms?"

After the conversation with Frink, Kevin set about finding another party already familiar with the specifications that could write the FFS2 code for us. He was close to signing with a company called Systemsoft when it abruptly withdrew its offer.

Kevin called Systemsoft to find out why. His contact there said, "Well, frankly, Microsoft has threatened to sue us if we help you."

Kevin stomped into my office with the news and waited for my response. "What blows me away is that Microsoft even knew about the proposed contract in the first place," I said.

Kevin narrowed his eyes and spoke softly. "Now I know what it's like to be blacklisted by the KGB."

Throughout the winter of 1992, the team hunkered down to meet a self-imposed deadline of the end of March to complete the first full release of Penpoint. For over a year, the only version available for sale had been the so-called developer's release, which was primarily intended for use by ISVs. It was a long and difficult haul, punctuated by persistent rumors of delays and problems. Overzealous engineers had burdened the design with arcane features of questionable value, complicating the development and testing process.

With more than a hundred programmers coding away at top speed, Robert's job was like choreographing a ballet for a centipede. Each morning, programmers would "check out" the software modules they needed and download them over the local area network. A special control system would then prevent anyone else from changing the same modules until they were checked back in that evening. All modules were due for return at a prescribed time, posted high on a wall. Then the buildmasters would take over — nocturnal trolls who pulled together the day's changes into a new, interim "build" for testing. We instituted "build dinners," so anyone who was around the office at seven o'clock could feast on programmers' ambrosia: a rotating carte du jour of gourmet pizza, Thai food, fried chicken, and pasta. As a show of solidarity with the engineers, not to mention the lure of free food, nearly everyone stayed late into the night. I gained five pounds that winter.

Robert offered to pay for baby sitters for those who needed to come in on weekends. We held a Family Appreciation Day, when spouses and children were invited in for presentations, to learn about why their loved ones spent so much time at the office. To help build morale, two senior technical managers with New York accents dubbed themselves the BBC — Brooklyn Broadcasting Company — and prepared a daily videotape of the day's progress, punctuated by important announcements for the team, spoofing network broadcasts of the Olympics.

The press, desperate for anything that looked vaguely like news about pen computing, amplified the slightest rumors to the status of gospel. Gossip columns had sprung up on the back pages of several trade magazines, and readers inevitably turned to them first. The editors caught on to this trend, but rather than dignify the gossip columns by moving them to the front, they shifted the important news to the back pages. Industry executives learned to read the trades from back to front, like Hebrew religious texts. When we settled on April 16 as the announcement date — mainly to avoid conflicts with some other big events scheduled for the end of March — *PC Week* ran an article labeling this a setback, suggesting that we were losing ground to Microsoft because of repeated delays.

The IBM negotiator that Randy had been dealing with continued to pressure him about the loan. Since our new financing had allowed us some breathing room, the negotiator decided to use our upcoming announcement as leverage to win rights to Penpoint's "look and feel," broadening his demands. "We need these rights to justify continuing with you as a partner," he would say. IBM was planning to unveil its own pen computer, cleverly dubbed the Thinkpad, at the announcement, but threatened to pull its support for the event up to the last minute.

"IBM isn't a partner, it's a competitor," Randy would retort.

The week before the announcement, Randy stopped by to pick my brain about the meaning of some particularly obscure language in the Penpoint license agreement with IBM. "Randy, they've got to be kidding. According to their interpretation, if we took a fifty-million-dollar one-time payment for a license from someone else, with no per-copy royalties, IBM's royalty rate would automatically go to zip. Besides, their public relations people are already out giving advance briefings to the press about the announcement."

Randy swallowed his distaste and proposed a face-saving resolution: we would consent to certain modifications to IBM's license, but the changes would go into effect only if the entire loan agreement was closed within forty-five days. Nevertheless, Randy and Bill had to spend the entire weekend in the office, working out the details. With only one day to spare, the negotiator finally put out the word inside IBM that everything was settled.

Just prior to the event, our partners began squabbling like children. The IBM people were annoyed because they weren't going to be the only hardware company to speak from the podium. NCR was miffed that we provided space for IBM to announce the Thinkpad. Vern Raburn complained that Slate wasn't getting enough play, despite the fact that Robert was going to demonstrate PenApps extensively. Editors at *PC Week* suggested that they might withhold front-page coverage because *InfoWorld*, whose office was nearby, got a chance to look at a Thinkpad before they did. Here we had rented most of the beautifully renovated Sheraton-Palace Hotel in San Francisco and spent more than half a million dollars to showcase the efforts of our partners, yet they were at each other's throats and blaming it on us. In contrast to the camaraderie of our earlier announcement, it was beginning to feel as if we were staging a brawl.

But then Robert walked proudly to the podium, surrounded by the large, pulsating "gestures" that dominated the stage set. "This is it," he said, gently lofting a shrink-wrapped copy of Penpoint above his head like a precious newborn. "Penpoint 1.0!" He spoke with passion and eloquence, and the crowd was spellbound. At our previous event he had to make do with demonstrating mostly GO's home-grown applications. But this time he had his pick of a number of creative applications unlike anything the audience had ever seen.

Robert debuted Numero, from Pen Magic, a reinterpretation of the paper ledger-sheet for a pen computer. He showed how you could fold over the electronic paper — by dragging in the margin with the pen — to put two columns next to each other for easy comparison. Drawing a line underneath a list of numbers caused a total to appear. After building a table of sales data in record time, he opened a "graph object" within a business letter and dropped the table into it, instantly forming a labeled pie chart. Writing an *L* on the chart changed it to a line graph. When he tapped on the graph three times, it became three-dimensional. The audience began stomping their feet in excitement. Concerned about the mayhem — which could be heard throughout the building — a hotel manager slipped in through a side door to monitor the proceedings.

When the house lights came up and the buffet tables opened, Stewart Alsop approached and shook my hand. "I've got to hand it to you. You guys have really convinced me this time. You haven't left much room for the keyboard bigots to argue."

After lunch, everyone went out to the broad promenade on the mezzanine, lined with booths where forty ISVs demonstrated their Penpoint products. That evening, we threw a party for the entire company at a local nightclub. Bill got up to make a toast, and was overcome with emotion. "Everyone thought it was vaporware. But today we really stood and delivered."

▣

"Christ. Last time it was Microsoft, this time it's Apple. Won't anybody cut us some slack?" Robert was wielding a disorganized handful of news clippings. Wayward columns of type flapped in the air as we walked down the hall toward Bill's office. Mike Homer was already there, constructing a table on the whiteboard comparing Newton with Penpoint.

As the reverberations from our announcement were dying down, Apple had deftly stepped in to fill the widening void. Penpoint 1.0 had been hailed as nothing less than the dawn of a new age in computing. The press coverage was so extensive that John Sculley decided he could wait no longer before throwing his hat into the pen-computing ring — he didn't want to look like an also-ran when the time came to introduce Newton. He had hoped to use the project to reestablish Apple as a technological leader. A master of promotion, Sculley knew that the secret to success was to change the name of the field and find a new audience. To my amazement, he was immediately successful. The fact that the release of the first Newton product was at least a year off proved to be only a minor inconvenience.

With little more than a videotape, a desktop demo, and a slim, one-pound model carved of wood, Sculley had arranged to be the keynote speaker at the Consumer Electronics Show, held in Chicago in May. The sponsors of this exposition — whose purpose was to showcase the latest in stereos, televisions, video games, and other assorted electronic flotsam — were pleased to have a senior executive "cross over" from the more business-oriented world of personal computers to address the assembled throng.

He regaled the crowd with optimistic predictions of "the mother of all markets" created by the convergence of computers and communications, which would foster a new class of products that he termed personal digital assistants (PDAs). They ate it up. Privately, though, many industry observers were concerned that his expectations for this new technology were overblown. "There's sure to be a backlash eventually," one of them cautioned, "and that's going to hurt everyone, not just Apple." After the obligatory editorials handwringing over the proliferation of acronyms, and a spate of columns attempting to define what PDAs really were, the term stuck. Suddenly computer companies were hiring high-priced consultants to formulate their PDA strategies.

"I guess the pen is not the point anymore," Vern Raburn started saying at conferences, taking potshots at GO's original positioning. I couldn't understand why he kept harping on this. Seemingly, it would hurt him as much as it hurt us.

I turned to Layne for advice. "Forget about Raburn, you can't count on him anymore," she said. "He doesn't burn bridges, he napalms them." As usual, she was right about such matters. From that point on, the relationship between Slate and GO deteriorated badly.

When we studied the press clips of the electronics show, something looked odd about one of the pictures, purportedly a prototype of the Newton.

"Take a close look at this," Mike Homer said to Robert.

Robert examined the picture closely. "I'll be damned. Those LCDs do tend to peel, don't they?"

The screen on the Newton was nothing more than a graphic printed on shiny paper and glued down. The corner had curled up from overuse.

A few days after Sculley's press blitz, I received an e-mail message from our manager of production for documentation. He had sent it to the entire company.

> I am leaving GO today. Unlike those who have preceded me, I am not going to another exciting startup (thank God). I am leaving to take care of my health. I have AIDS. My purpose in telling you this is educational. While I am sure that some of you have had direct experience with this, I also think that for some of you, AIDS remains a distant event happening outside of your lives . . .
>
> None of us really know how long we have, despite the cultural denial of mortality. So, I urge you to take a moment and think about what is REALLY important to you and then put some extra time and energy toward it . . .
>
> So that's it. It has been a pleasure and a privilege to work with this great team. I will be avidly following GO's meteoric rise to success.

Fourteen months later, he lost his fight with AIDS.

Like some cave-dwelling species of salamander, our struggle for survival was so relentless that the seasons slipped by unnoticed. Those of us fortunate enough to have offices with windows could tell it was summer by the angle of the sun on our computer screens, bleaching out the words with its hot glare. For the rank and file in their catacomb of modular cubes, the natural world was even less intrusive. In an attempt to encourage a more healthy balance, Bill Campbell invited the executive team over to his house for a Saturday afternoon of fun, determined to get in some token relaxation. Besides, it was about time for a complete strategy review.

Bill's large colonial-revival house sat on a full, flat acre in prestigious Old Palo Alto. He had converted the top floor into a game room, with billiards, pinball machines, darts, workout equipment, and a modest bar. After a few games of pool, we settled into some lawn chairs on the brick patio outside the kitchen. I chose a spot under the awning.

"Kaplan, aren't you going to sit in the sun?" A sign of the season,

Bill chose to use the standard summer-camp appellative form: the unadorned last name.

"He has to maintain his waxy nerd complexion," Randy said before I could answer.

"Yeah, he stocks his sunlamp with Cool-White fluorescent lights," Mike Homer added.

"I guess I'm just a cool white sort of guy," I said, putting on my shades and smiling. "By the way, homeboy, how come you're always double scheduled on meetings? You keeping two sets of books?"

"That's only an illusion," Mike said. "It's the Doppler effect. The meetings get all compressed on the calendar as they approach and then sort of spread out as they pass by."

Bill got down to business. "Well, I guess we all know the good news."

Randy clapped his hands together. "We're all fired?"

"Not yet," Bill said. "We finally won the AT&T Telepen RFP — by the skin of our teeth. Boy, was Porat ticked off." Marc Porat was the preppy CEO of General Magic, the Apple spinoff mainly developing a new communications protocol. He wasn't accustomed to losing. "I heard he was wandering around the halls in Basking Ridge, collaring any AT&T executive who would listen to him." Mike shook his head in mock sympathy.

Bill picked up the full bottle of light beer he had been pretending to drink all afternoon. "So we're really at a crossroads. I think it's time to shift our top priority from IBM to AT&T. Just in case anyone disagrees, you might want to hear this. I had lunch with the CEO of Communications Intelligence Corporation last week. He told me that IBM has licensed their DOS-based pen software and is planning a big push with the ISVs." Everyone looked stunned. "But he told me they don't want to announce until some deadline with us passes."

"The loan agreement," Randy said. "They probably don't want us to find out until after we sign the deal. Yet another well-kept IBM secret."

"The other good news," Bill said, "is that we're finally ready to close the IBM loan. But before we do it, I want you to hear what Randy has to say."

Randy took his feet off the edge of the patio table and sat straight up. Now he was dead serious. "I wouldn't exactly call the loan good news. We have to do this because we need the money. But it includes one term that I predict will eventually cause us real trouble. If any other party wants to buy more than fifteen percent of GO, we have to offer the same deal to IBM first. That's the easy part. But if IBM declines, then they have the right to call the loan."

"Ugh," I said. "That puts a ten-million-dollar-plus-interest dowry on us if we decide to get married to someone else."

"Exactly," said Randy. Here it was July already and GO was running low on cash again, not to mention that we were all tired of negotiating with IBM. But by that point, IBM was in the process of imploding. Racked by growing losses, wave after wave of lifetime employees were taking early retirement, accepting reassignments, or just plain quitting. It was amazing that IBM was proceeding with the loan at all.

The best people — those who were the most able to find new jobs — were the first to leave. Sue King, our only effective internal supporter, had already jumped ship to join Apple, leaving in her wake the half-finished Thinkpad project without a forceful leader. The ones who remained behind — like foot soldiers abandoned on a decimated battlefield — began building a case that we, not they, were responsible for the probable failure of their project. The acclaimed product release wasn't followed through, and the Thinkpad became available only by special order. This meant you practically had to beg them to sell it to you, if you could find someone who had even heard of the product. By the time the first prototypes were delivered, the Thinkpad had earned a reputation for being too heavy (over five pounds) and too expensive (over $5,000).

Robert was nervously squeezing his empty can of Diet Coke, slowly reducing it to a crumpled mass of sharp points and raw edges. "Well, then, let's just take the money and bid them good riddance."

Randy looked around. No one else wanted to comment. "Fine. Bill, when the time comes, are you going to call IBM and tell them their project is no longer our top priority?"

"Frankly, I wouldn't even know who to call anymore."

Within weeks of our one-day retreat at Bill's house, even Kathy Vieth had left — the last responsible executive who was moderately familiar with the GO–IBM relationship. With the loan deal completed, our contacts with IBM faded away, as though we were different continents imperceptibly drifting farther apart until the last land bridge sank into the sea. All that remained were a number of contractual time bombs, buried deep in a legal department file cabinet, ticking away.

Based on our experience with IBM, everyone knew that our first order of business was to have AT&T go public with its support of GO as quickly as possible. Before that happened, there was always the possibility that some senior executive could turn around and decree that another party should win the Telepen RFP. The peculiar thing about the Request for Proposal was that it really was no contract at all, merely an internal AT&T understanding that the winner would become involved in AT&T's future products. Bill referred to the RFP as a "license to hunt and fish."

Mike Homer, as our VP of marketing, saw that the first chance for a public announcement was at the upcoming Mobile92 conference, put on by the industry analyst and ex–*Wall Street Journal* reporter Dick Shaffer. Shaffer's organization, Technologic Partners, published a newsletter and ran conferences targeted to computer industry investors, analysts, and executives. Jumping on the pen-computing bandwagon, Shaffer had decided to start a new annual conference, to be held in mid-July at a hotel near the San Francisco airport. Whereas most conferences take a few years to develop a constituency broad enough to make them profitable, this one was an instant success — it attracted thirty-nine exhibitors and more than eight hundred attendees.

The big draw at Mobile92 was Larry Tesler of Apple, who for the first time was going to demonstrate publicly a working prototype of Newton's software. He was supposedly taking time off from a sabbatical to make the presentation, but I later heard that he had actually been sequestered at Apple, working night and day on future Newton plans.

Mike was determined to blunt the effects of Apple's public relations blitz in general, and Tesler's presentation in particular, in any way possible. Working with the AT&T PR department, Mike attempted to rename and reposition the field. He developed a lengthy briefing paper about a new class of devices called personal communicators, made possible by the joining of Hobbit and Penpoint technologies. The piece contained graphs, predictions, and sketches of futuristic devices combining the functions of pen computers and mobile telephones. He arranged to have a copy of the briefing paper placed on every seat in the hotel's ballroom on the afternoon Tesler was to speak. Then he scheduled a joint press conference with AT&T for the morning of Tesler's talk. It was a bold and brilliant marketing move, designed both to cement AT&T's support and to excite the public.

Meanwhile, Bill had cultivated a personal relationship with Bill Warwick, a seasoned and effective AT&T executive who had the ear of the CEO and whose current assignment was to turn around the microelectronics division. AT&T Microelectronics, the $2-billion-a-year integrated-circuit design and manufacturing arm of the phone company, was responsible for the Hobbit chip. Though it was one of the largest chip manufacturers in the world, it represented only about 3 percent of AT&T's revenue.

Warwick, who was at least ten years Bill's senior, had the relaxed, affable manner of a television grandfather. Fortunately, he was willing to put the full force of AT&T's credibility and support squarely behind GO and Penpoint, which he did at the joint press conference. He even announced the creation of a new business unit, AT&T Personal Communications Systems, to develop, license, and sell products based on the Hobbit-Penpoint technologies. "GO is the leading provider of mobile operating-system software," he intoned with a sonorous voice. "We look forward to working closely with our new partner to develop a powerful new standard platform for personal communicators." He then presented slides describing an elaborate and expensive ten-year development plan, culminating in the production of wireless, portable, full-motion video phones.

Then Bill got up to speak. "We have one advantage over AT&T in this relationship." Everyone perked up at this odd comment. "Our founders are still alive!" Following his presentation, and overcome with gratitude at Warwick's unflinching support, Bill walked over and gave him a big hug. Warwick's face turned a deep red at this show of emotion.

"This is what IBM was supposed to do for us," Bill said to me privately as we walked out together. "Can you imagine Cannavino ever saying those things? Warwick is a stud buffalo in my book." He was beaming.

That afternoon, Robert and I sat together throughout Tesler's demo. The software ran on some sort of prototype board mounted in a Macintosh, but nonetheless it was fabulous. Newton had the trademark Apple playfulness — little sounds and clever animations — that showed off the true power of the ARM chip. Apple's engineers had plainly learned whatever they could from GO and others, but had the intellectual honesty to develop their own distinctive user interface. For the five years since the original meetings with Steve Sakoman in a Boston hotel, the engineers had been able to focus all their attention on building a great product, undistracted by hardware partners, fundraising, and customers. And now they were ready to show their results. Robert and I were impressed — and more than a little worried.

When Tesler finished to thunderous applause, Robert leaned over to me and whispered, "One thing's for sure. This shows that there are lots of ways to do it — Microsoft didn't have to copy from us. But we better hurry the hell up and get our Hobbit version out with EO."

The strong AT&T announcement counterbalanced the Newton announcement somewhat, but we all knew we had barely dodged another bullet. "Looks like we made the switch to AT&T just in the nick of time," Kevin said at the conference reception. "Without them, we would have been buried alive."

The Bubble

WITH Microsoft, IBM, NCR, Grid, Apple, AT&T, and GO all trumpeting their efforts at every opportunity, a firestorm of publicity raged throughout the last half of 1992. The hype had gone well beyond the trade magazines: GO was featured on a CNN piece spotlighting companies that were thriving despite the recession; *USA Today* ran an article on the front page of its Money section, with a large color photo of me showing off Penpoint. The weekly pile of press reports provided by a professional clipping service had grown so thick that no one had time even to scan it all. Any conference with "pen" in the title drew huge crowds of curious onlookers wondering what all the fuss was about, and most of them left impressed — by the size of the crowds, if nothing else. The smoke was so thick it didn't seem to matter that there was virtually no fire at all.

In this overheated environment, I was unexpectedly invited to speak at a technical seminar at Intel, to help educate its marketing staff about the potential impact of PDAs. There were six participants, representing hardware and software companies doing work in pen computing — including Microsoft, which sent Lloyd Frink, the technical leader of the Pen Windows project. Intel had asked Michael Rogers, a technology writer for *Newsweek*, to be the moderator.

The panel convened in a large, windowless, amphitheater-style classroom, with a projection booth at the back and a narrow stage in front. It was a beautiful day outside, so the ventilation system had to work that much harder to remove any trace of the sweet, lazy afternoon air of August. We sat on the stage behind a row of long tables littered with tented name cards, plastic jugs of ice water, and complimentary leather folios with "Intel" stamped on them in gold leaf. Tacked on the wall to our right were large framed copies of Intel's corporate mission statement, in case anyone wanted to brush up should the presentations drag. To my surprise, the room was packed, despite the fact that it was a warm August day.

Rogers started off by referring to a recent comment — attributed to Andy Grove, Intel's CEO — that the PDA market was little more than "a pipe dream driven by greed." He followed this with some tongue-in-cheek remarks lamenting the proliferation of acronyms for the new field. Then the panelists each spent about fifteen minutes presenting their points of view, mainly in vacuous generalities. Given the size of the panel and the participants' inability to stay within their allotted time, these preliminary statements took up close to two hours, during which portions of the audience seemed to nod off.

Finally, Rogers asked each panelist to answer the question "What can Intel do for me?" By this time, even the speakers were looking at their watches and hoping someone would pull the fire alarm. When he got around to me, I dutifully hit on the subject of not designing future microprocessors around DOS. "Don't make us pay for DOS compatibility," I said, a subtle jab at Microsoft's licensing practices. The suggestion that DOS might not be required for a pen computer was a novel concept to much of the audience, since the ubiquity of DOS on Intel processors was precisely what was paying their mortgages. The remaining panelists were happy to have a real issue to talk about, so much of the subsequent discussion centered around this question of desktop compatibility.

Rogers toyed absent-mindedly with a pencil as he gave Lloyd Frink his chance to speak. Throughout most of the afternoon, Frink had sat quietly hunched over, staring down at the table in front of him as though lost in thought. Now he looked directly ahead and spoke as if in a trance, addressing no one in particular. "There are two kinds of compatibility. One is being able to leverage the development tools, which we think is important, and the other is the applications — being able to use the same applications." Everyone was waiting for him to extol the virtues of leveraging the installed base of DOS and Windows applications, which of course required an Intel-compatible processor. Instead he said, "And, um, we don't think that's important for this market."

Those who were still listening looked around the room, then quickly roused their neighbors. Could this really be what Frink meant to say? As a spokesman for Microsoft, could he be suggesting that pen computers were not going to be mere extensions of the desktop monopoly that had built the twin powerhouses of Intel and Microsoft? Was he really repudiating the basic premise behind Pen Windows — that compatibility with keyboard-driven computers would be enough to steamroller any competitive challenges?

Rogers dropped his pencil to the table. The entire audience now paid close attention. Oblivious of the abrupt change of mood, Frink continued. "It's going to require new applications, a new interface, much easier to use — simpler than what you even get with Windows today." Everyone froze in silence at this confirmation of their worst nightmare, brought to life by the only person whose credibility and lack of self-interest in making the statement could not be questioned. A wall thermostat released a hissing sigh, telling the overhead vents to drop a fresh blanket of cool air on the room.

Still staring straight ahead into the distance, Frink simply blinked once. It wasn't the reflexive blink that clears the eye. Nor was it the subtle, deliberate blink that a gambler uses to signal his partner. It was the lingering, involuntary blink of capitulation, of reluctant acknowledgment of a change of heart. In that pure silent instant, which swung past like the flash of a distant lighthouse beacon, I knew that we had at last made our point. Five years of hard work, the efforts of an army of dedicated supporters, and over $50 million, all came down to this one extraordinary moment. Frink's modest disclosure of Microsoft's new direction was the final stroke needed to transform opinion into truth: pen computers *were* something new and different, a class of device distinct from keyboard-driven machines. From that moment on, I never again heard this proposition credibly questioned.

When the seminar ended, Lloyd Frink and I left without exchanging a word. I drove north, back to my office. When I pulled into the parking lot, Robert was just getting into his car. "So, how'd the panel go at Intel?" he asked facetiously. "Who'd they believe, us or Microsoft?"

"They blinked," I said.

"Who blinked?"

"Microsoft."

He looked wistful. "Huh" was all he said. He reached out of his car window and shook my hand. Then he added, "It's about time, my friend."

A principle widely held by physicists states that there is no such thing as a truly independent observer, that every act of perception, no matter how trivial, affects whatever is observed. Information theorists recognize a similar principle: every message, no matter how perfunctory, is carried by a medium that colors its meaning for the recipient. An Egyptian obelisk, standing tall in the shifting sands, tells of gods and kings with permanence and power. A tightly folded classroom note, surreptitiously passed from desk to desk, speaks with urgency and danger. And electronic mail, dashing through the netherworld of packet-switched networks, delivers its data with indifference and harshness. That's why Bill Campbell, the embodiment of warmth and charisma, almost never sends e-mail.

So the hallways buzzed with rumors when Bill electronically ordered the entire company to attend a meeting at noon on Monday, August 10, 1992. It added to the mystery that his summons was followed by a series of codicils from his administrative assistant, amending the time to and fro until it finally settled at one o'clock. Up to the moment Bill stood to address the crowd, the staff spent the last hour monitoring their computer screens for further postings, forestalling lunch, once again proving that fear is stronger than hunger.

The employees gathering in the main conference room remained tense until Bill broke into a proud smile. "I've called you here today for a special reason. We have a distinguished visitor, whose name you may have heard around our halls for some time. He has just arrived by corporate jet from New Jersey to talk to us about our partnership and his vision for our future together. I'd like to introduce Bob Kavner of AT&T." The audience could tell from Bill's uncharacteristic formality and dramatic tone that he was impressed by Kavner's willingness to make a personal appearance. Indeed, no IBM executive of any significant rank had ever taken the time to speak at a company function.

Kavner's background was as an accountant, having spent eighteen years with Coopers & Lybrand. After joining AT&T as its chief financial officer in 1984, he worked his way up the organization chart. Ultimately he headed the multimedia products and services group, reporting directly to Bob Allen, the company's CEO.

Since Kavner's plane had been delayed, Bill didn't want to postpone the meeting any longer and so didn't take the time to brief him adequately. As a result, Kavner was unclear about what Bill wanted him to say. Heads strained to get a good look as Kavner stepped forward to speak. A pale-skinned man with a prominent nose and a crest of gray hair, he looked like a photograph of a parrot shot in black-and-white. He stood motionless as he delivered his remarks, making him appear uncomfortable. His reserved manner contrasted sharply with Bill's barely restrained exuberance.

"You are all to be congratulated on winning the AT&T Telepen RFP. You had some stiff competition, but were selected nonetheless." Kavner's eyes darted around the room when he spoke, as though he were looking for a familiar face. "But I want you to know that this is just the first step. Lots of these projects never lead to any revenue for the companies involved. Now you have to deliver a quality product, on schedule."

Bill's smile began to fade, betraying his disappointment. The words were there, but this was not the enthusiastic salute and inspirational call to arms that he had expected.

"One more thing," Kavner said, turning to address Bill as much as the audience. "It's important that you attract some other strong partners. The more big players behind you, the better for everyone. You shouldn't be counting on AT&T as your only partner."

Did Kavner mean that AT&T wasn't really committed, after all the hard work and public pronouncements, or merely that he would welcome the addition of other parties to help build momentum? His statement was ambiguous at best.

Bill tried not to look concerned. He abruptly ended the meeting without permitting any questions, using the delayed lunch hour as a convenient excuse to move people gently on their way. The crowd shuffled out, muttering among themselves about what Kavner might have meant.

◲

I tapped the side of my glass with a fork to get everybody's attention. The caterer stuck her head out of the kitchen, on the off chance that I was trying to summon her, then disappeared again. "Dearly beloved," I intoned, "we are gathered here today to honor one of our own, the inimitable Kevin Doren, burner of the midnight oil, keeper of the sacred EPROM, and defender of the pen faith."

Kevin pounded the table, cutting through the laughter. "I'm not dead, you know, just leaving." The board of directors and senior executives — those of us who were in town — had gathered at my house for his going-away dinner. Since the EO spinoff, there had been progressively less need for Kevin's skills, and he was mature enough to realize that even as a company founder, it was best for both him and the organization to accept this sad fact and take appropriate action. A few weeks earlier, Bill and Kevin had agreed that he should move on sometime in the next few months, and now the time had arrived. Robert and I both felt as though a brother-in-arms, decorated repeatedly for bravery in combat, was being recalled from the front and returned to civilian life.

After several bottles of Napa Valley chardonnay and a dinner of pasta primavera, the mood was convivial, if not a bit rowdy. We went around the room telling our favorite Kevin stories, which were usually about his miraculous gift for fixing complex electronic equipment with ordinary household articles such as toothpicks, straight pins, and rubber bands. After he was thoroughly roasted, the conversation turned to more serious matters.

"What do you think about the problems with EO?" Robert asked David Liddle.

Before David could respond, Randy Komisar jumped in. "The problem isn't EO, it's that Rossmann never sent over the engineers he promised to help with the Hobbit port."

Several months earlier, Bernie Lacroute had stepped down as EO's chief executive, having completed his promised year of service, and took the title of chairman of the board. After searching outside the company for a replacement, the board appointed Alain Rossmann, EO's vice president of marketing, to the position. Since then, Rossmann had shown himself to be undiplomatic and hard-nosed in his dealings with GO and other business partners. Although Bill Campbell had originally supported Rossmann, their relationship had begun to sour following his appointment. As Bill was now in Japan, and so was not present at Kevin's dinner, this seemed to Robert to be a good opportunity to elicit some independent opinions.

Under Rossmann's watch, the resentment toward GO that had unfortunately been sown at EO's inception continued to grow, creating an attitude of isolation and a determination to go it alone. GO had spent years building relationships with its ISVs, customers, and partners, and now many of them found EO less cooperative and supportive of their efforts.

Even worse, AT&T provided EO with extensive financing — tens of millions of dollars, more than the company required for its core hardware business — because they saw EO as a source of tangible products that could be marketed under an AT&T label. Rather than investing this money exclusively in hardware development, however, Rossmann undertook to differentiate EO's offerings from those of other Penpoint vendors through software, by building or buying applications to bundle with his products. He sometimes sought to make these arrangements exclusive, meaning that no other Penpoint vendors could offer the same applications to their customers without a license from EO, a competitor.

From Rossmann's perspective, this was just good business. But it had the side effect of discouraging other hardware manufacturers from working with Penpoint, because they appeared to be at a permanent disadvantage to EO. This strategy might lead to short-term gains for EO, but would hurt everyone, particularly GO, in the long run.

David prevented the discussion from descending into a bitching session. "This is a real strategic problem with no obvious solution," he said. "The thing is, Rossmann is closer than we are to AT&T. We have to find a way to prevent him from drilling holes in the Penpoint boat below the water line before we all sink." He turned to

John Doerr, who had been oddly subdued during the discussion. "Why don't you sound Bernie out about working with Rossmann on his attitude, maybe even recommend that he step aside?"

John heaved a deep sigh. "I already have, but the EO board is happy with Rossmann. There's nothing more I can do."

Just then the kitchen door swung open and dessert arrived. As everyone else formulated a plan of attack on their chocolate soufflés with crème fraîche, I fetched a speaker phone from my home office and plugged it in. "You guys have to hear this voice mail message. Remember the article about pen computing a few weeks ago in the *Times?* Well, get this. My mother calls me, and she's indignant that my name is spelled wrong. Hey, what good are bragging rights in Palm Beach if the *New York Times* misspells your son's name? So I suggested that she call up John Markoff and make sure he knows the correct spelling. I figured she'd never do it. Anyway, a few days later, I get this on my voice mail."

I dialed up the office and played the message. "*Jerry, this is John Markoff of the San Francisco bureau of the* New York Times. *I just wanted to let you know that I will never, ever, misspell your name again, so help me God. J-e-r-r-o-l-d.*"

"Hey, we should hire her in the PR department," Robert said. He narrowed his eyes, pointed his finger at the phone, and said in a scolding mother's tone, "Don't you *dare* talk that way about my son's company! He's a good boy. Always cleans his plate!"

After dessert, the guests dribbled slowly out, first the ones with kids, then those with wives. But Kevin stayed until last. "You know, Jerry, in a way this is a relief."

I put my arm on his shoulder. "I can't even imagine it. You look two inches taller already with all that pressure off."

"But I want you to know that I don't regret one day of it. It was just like you promised — one hell of a joyride. I'm just sorry I won't be there when we reach our destination. And now with AT&T as a partner, we're almost there."

"Yep. You had the good sense to get off one stop before hell."

We both laughed. Neither of us suspected how true this statement would turn out to be.

A few days later, I met with Brook Byers, one of the more senior partners of Kleiner Perkins Caufield & Byers, about GO's future financing needs.

Byers shook my hand. "Man, what happened to your hair?"

Since we last saw each other, I had gone almost completely gray. "One gray hair for each day as an entrepreneur," I said. He looked much the same as the first day we met, five years ago.

Byers listened sympathetically as I explained my view of the market, where we stood, and the still excellent prospects for success if we could just see it through.

"Jerry, you should know that there's no way we can let this thing go south. KP's got too much at stake." I assumed he was talking about losing the money, but I was wrong — he was at least as concerned about the firm's reputation. "You've got this pen thing stoked up white hot. One way or another, every firm in town has an investment in a Penpoint-related venture. If you don't succeed, it's not going to look very good for any of us. Right now AT&T's the only deep pocket in sight. That's a big problem."

"How come?"

"In finance, you always have to have an alternative," Byers said. "Otherwise, you're sure to be eaten alive."

"Even with a blue-chip company like AT&T?"

"With any large company. That's why I recommend that you look into a Reg D with a retail investment bank."

"What's a Reg D?"

"Regulation D allows you to sell stock to individual investors of high net worth without having to register with the SEC as a public company. In a matter of weeks, retail brokers can sell GO stock in fifty or hundred K chunks to high-net-worth individuals. The bankers will take a big fee, of course, but they can move it out in no time, because their investors have made lots of money on these deals in the past, and the brokers get warrants on the company's stock as an incentive. Reg D's have been very popular in the biotech area."

The way Byers talked about it made it sound pretty simple. "But what makes you think they'd do it for us?"

"It's a hot market, and they're hungry for product. But you better hurry up. I doubt that the market will hold much past New Year's.

They can pull together twenty-five to fifty million at a very attractive price. You should give them a call."

My eyes lit up. "Twenty-five to fifty mil at a high price?" He didn't have to ask twice. As I drove away, I wondered if there was any limit to how high we could fly with that much capital behind us.

◻

"COMDEX, COMDEX, COMDEX. I can't believe it's COMDEX time again!" Robert said. We stood amid a scene out of a nightmare carnival. The red-and-white-striped registration tent seemed to span several acres, and was almost entirely filled with rows of those ribbon dividers used at banks and airports to channel people into one enormous snaking line. After half an hour, we were about midway through, having zigzagged perhaps twenty times. A sympathetic woman in a gray smock was stationed in the middle, passing out small paper cups of ice water from a large Styrofoam container. "It's about fifteen minutes from here to where they make the name badges," she announced.

"Either the years are getting shorter or I'm getting slower," Robert said. "It seems like last year's show was only a few months ago."

"Twelve, to be exact," I responded. I wasn't in the best of moods. But he was right. Somehow the world seemed to be spinning faster and faster, like a carousel run amok. I sometimes forgot which hotel I was staying in, confusing one year with another. My only island of stability was the new life I was trying to build with my wife, Layne. But except for a long honeymoon weekend at Vinod Khosla's vacation home in Carmel, we could find very little quiet time to spend together. I was fortunate that her background was also in the computer business, so at least she understood the demands on my time, but there's a limit to what anyone should be expected to put up with. I knew that eventually something had to give, and despite my father's advice that work comes first, I was determined that it not be my marriage.

When we finally got our passes, we rushed over to see the AT&T exhibit in the main hall. Word had spread rapidly through the Penpoint community that this COMDEX was going to be our big break. EO had just assembled a batch of their first product, the EO 440, and it was awesome. It had more power than a fast 386, packed into a slim, two-pound package about seven by ten inches and less than an inch thick. (The nearest competitor, running Pen Windows, was about twice the weight and size.) On either side were two semicircular "ears," one containing a microphone and the other a speaker. An optional flat cellular-phone module could be tucked underneath, with a cradle for a handset at the top, so you could make phone calls and send faxes while on the run.

AT&T had spent a fortune on a large booth to showcase the new technology, adorned with colorful flags and posters, and offered it to EO, gratis. The center of the space was an open theater with seating for several hundred. Eight-foot panels stood at irregular angles, displaying large billboards with a schedule of presentation times. Between shows, lively music played as a small army of EO, GO, and AT&T demonstrators milled about the area proudly showing Penpoint on the EO 440. Their loud voices and black-and-white-striped jerseys made them seem like hawkers at a sideshow. I watched as visitors darted about from demo to demo, straining to get a closer look.

Someone tapped me on the shoulder. "You look pretty dapper in that fancy business suit, Jerry." It was Jim Cannavino of IBM, with his personal assistant and several other associates in tow.

"Well, we're a small company, so I have to dress like I come from a big one," I said. Then I noticed they were all wearing khaki pants and pastel sport shirts open at the collar. Apparently they felt that this informal uniform would reinforce the image of a new, easygoing IBM. "And I guess since you're from a big company," I said, "you have to dress like you're from a small one." Cannavino was relaxed and friendly as we chatted in the booth of his major competitor, AT&T, which was heavily promoting GO's product. He seemed to be unaware of the fading relationship between our two companies, or at least he chose not to acknowledge it.

On Monday of that week, AT&T staged an announcement of their "personal communicator strategy" to an audience of a thousand or more. This production, in the big theater at Bally's, was by far the largest and most expensive demonstration of GO's technology. Robert's demo of Penpoint easily overshadowed the prepared speeches and flashy videos presented by a sequence of AT&T executives. Then Alain Rossmann showed how a fax could be received on the 440 through a wireless cellular connection, to the delight of the crowd.

The announcement proved to be the main event at COMDEX. For the rest of the week, everyone was talking about the EO 440 and Penpoint. For the first time, it seemed that we had broken through and successfully delivered our message to the broader COMDEX community. Microsoft was relatively subdued by comparison. At various panels, Vern Raburn of Slate still promoted his position that the operating system didn't matter. He had cut a deal with Grid to bundle his Pen Windows applications with Grid's new Convertible, a cleverly designed four-pound package that was both a full laptop and a pen computer. But despite his efforts, the product was destined to die after an initial spurt of sales to the curious. Personal communicators were the happening thing, and since Raburn had ceased all Penpoint development for several months, he was distinctly behind. Apple, which usually kept a low profile at COMDEX, showed an inanimate model of Newton under a Plexiglas cover in a room off the main hall, but it attracted modest attention.

After a full week of press interviews, demos, customer meetings, analysts' breakfasts, panels, ISV briefings, and parties, Robert and I boarded the plane back to San Francisco. "I'm so tired, I feel like I'm having an out-of-body experience," he said.

"Mine's more like an out-of-mind experience."

As we took off, he looked out the window and smiled. "Well, I think we're finally on our way. If only IBM had gotten behind us years ago, the way AT&T just did."

I was a bit less sanguine. "Don't pop the cork yet. EO is a long way from shipping anything. We haven't even delivered final code to them, and it looks like we still have serious problems with the size of the code. Let's just pray they get their product working and out the door before the echo dies down."

He laid his head back against the seat and sighed. "Sometimes I feel like just running away and joining the circus."

We looked at each other, realizing this was the perfect setup line. I didn't have to say it. We both burst out laughing.

Soon after COMDEX, we held a board meeting to discuss our upcoming need for cash. In contrast to previous rounds, this time everyone was optimistic about our ability to attract the funds. The growing public awareness of pen computing, which we were instrumental in feeding, made investors more willing to open their wallets. Nothing strokes an investor's ego more than having a friend or colleague mention a high-flying company in a hot new field and then pointing out that he's in on the action. The press was enamored of us yet again, and we were about to ship Penpoint J, a more advanced version of Penpoint for the Japanese market. We hoped to follow this up, by the middle of the next year, with a similar version in English. Our new VP of sales had delivered a rosy forecast of 93,000 units in worldwide sales for calendar year 1993, backed up with detailed customer-by-customer and country-by-country projections.

But privately, the executives were uneasy. GO was over five years old and we were still running mostly on promises, like greyhounds chasing a mechanical rabbit around a track. The team was showing signs of wear. Engineers complained to each other of burnout. A growing number of employees had earned out their original stock grants. There was a lot of talk about the good old days, which never really existed. In contrast to the usual banter at our estaff meetings, the jokes were mostly gallows humor followed by nervous laughter. The plain fact was that startups aren't supposed to make it this far without turning a profit. But somehow we were still in the race.

Randy Komisar had adroitly secured a verbal commitment for a Reg D offering at the sky-high price of $4 a share, despite the fact that the miniature speaker in his trick Halloween socks had delivered a ghostly howl when he bumped his foot during a formal presentation. Now GO's outside directors were discussing the details of the offer.

"What's that after the commission?" John Doerr asked. He strained across the table to reach a cookie from a plate sitting next to a bowl of grapes that had been nearly picked clean.

"About three-sixty, three sixty-five a share," Randy said. "That's a post-money valuation of one hundred fifty to one hundred seventy-five million, depending on how much we raise."

I was worried that this price was unrealistic, and that the offering might not fly when it hit the street. "I just can't see how our business plan supports that valuation," I said. "It's enough to make a sane investor toss his cookies."

John couldn't resist flinging his macaroon at me in protest. "There's no right price, Jerry. It's just willing buyer meets willing seller."

"What about AT&T?" Vinod Khosla asked, changing the subject.

"I talked to Atkinson last week," Bill Campbell said, "just to let him know we're considering the Reg D. He was concerned that we not get too many investors to answer to, in case AT&T ultimately decides they want to take a big position in us."

Vinod's eyes narrowed. "You should try to nail him down as soon as possible. The window for IPOs could close at any time, and the Reg D would go with it."

"I agree," said Bill. "What I want to do is give him a deadline on a ten-million-dollar AT&T commitment. According to our business plan, that should be enough to see us through to profitability — if our sales projections are in the ballpark and we get the next version out on time. I'll tell Atkinson he has until mid-December for a go, no-go decision."

"So to speak," Mike Homer added. He was picking the last good grapes out of the bowl. "It seems to me this could put us on a path to an acquisition by AT&T. Are we sure that's what we want to do? Personally, working for AT&T isn't what I signed up for. Maybe we should do the Reg D to remain independent, and go for an IPO instead."

"Not a chance," said Bill. "Securing AT&T's continuing support is essential. We aren't going to mess with them while they're courting us, so we're all just going to have to be patient, whether we like it or not."

Mike grinned at Bill. "I can't imagine you reporting to Kavner!"

Bill looked him right in the eye. "I won't be. He'll want to put his own flunky in here to run the show." The room fell into an awkward silence. Never before had Bill suggested that there might be an end to his reign as GO's CEO; his boundless energy and undying loyalty to the team seemed eternal. The mere suggestion of an end, even if in jest, cast a shadow over the meeting.

David Liddle continued the conversation. "We're not a good candidate for an IPO anyway. First of all, the financials aren't going be there for a long, long time. Second, if we're important enough to

AT&T to justify an IPO to the underwriters, they aren't going to let us do it."

"Why not?" I asked.

"It's the 'bubble instability.'" Everyone stared at him, ready for some dazzling metaphor. We weren't disappointed. "Remember blowing soap bubbles as a kid? You know how sometimes you get a great big bubble with a little bubble stuck on the side? It stays that way for a while, but then they always either break into two separate bubbles or combine into one." He looked around the room. No one had a clue what he meant. "Well, it's the same thing with corporate partnerships. Either the little company doesn't matter to the big one — in which case they inevitably wind up going their separate ways — or it's real important, in which case the big company can't take the risk of not owning the little one outright. There's no in between."

Try as we might, no one could come up with a counterexample. David's insight was startling in its simplicity and evident truth. This was the reason virtually all of the "corporate partnering" arrangements between small companies and large ones don't work out. Either they separate — usually wounding the small company fatally — or the smaller one gets absorbed.

I was the first to speak. "You're implying that if we're important enough to AT&T, they'll insist on owning us?"

"That's right," David said. "Just wait and see."

It was a sobering prediction that no one in the room could easily accept. Still, it was clear that the right path was to turn down the Reg D and take the AT&T money if they offered it, which they soon did.

The Reversal

THROUGHOUT the winter of 1993, relations with EO deteriorated. Rather than work with GO to help build the overall Penpoint market, Alain Rossmann continued to pursue his own company's independent interests. He chartered his software group to develop additional APIs for Penpoint, a prerogative that most operating-system vendors guarded jealously.

According to the EO-Penpoint license agreement that Randy had negotiated at the company's founding, EO was expected to send several engineers over to GO, to help port Penpoint to the Hobbit chip. A final $3 million was also due to GO, portioned out at various milestones during the development. Despite regular requests from Robert, Mike Ouye never sent over the promised help, claiming that his engineers were tied up with other matters, and insisting that if they did come, they would work only on projects of immediate value to EO. Since GO had borne all of the costs of the port, and both companies had agreed to certain revisions in the development plan, Bill and Randy believed that the $3 million was now due and payable.

Rossmann disagreed, of course, and a protracted negotiation began. It was still in full swing when the time came to deliver the final Penpoint code to EO, so Randy naturally suggested that he might not send over the software until EO paid up. Although Rossmann had agreed in principle to the settlement, he was badly irritated by Randy's tactic. This, in turn, fed Bill's suspicions that Rossmann might not follow through.

John Zeisler, who had recently replaced Michael Baum at Pensoft, decided to play peacemaker by taking both CEOs to dinner, at a trendy sushi bar in Palo Alto one evening in February. As long as Zeisler was present, things were civil enough. Unfortunately, the other two men had parked their cars near each other a few blocks from the restaurant. After dinner, they chatted tensely as they walked down a quiet side street.

Bill described the following scene at the estaff meeting the next day: "Rossmann's about to get into his car. I tell him it's obvious that he owes us the money, and I want him to know that we're not going to deliver the source code until he pays us what he owes us. Then he blows his stack."

Bill gestured in the air with his hands, imitating Rossmann. "You're threatening me! Don't you threaten me! I'll destroy Penpoint! It won't be worth a shit."

Bill returned to his own voice. "I couldn't believe that Rossmann would throw a fit right there on the street. So I figure I better take the high road." Bill looked toward the window as though staring right at Rossmann, and repeated his words from the night before. "It sounds like you're the one that's doing the threatening, Alain. I have no intention of threatening you. This is a business deal. You violated the business deal, you owe us the money. I'm not going to stand here listening to you scream your fuckin' lungs out."

The story finished, Bill looked at each of us and chuckled nervously. We all knew that as GO's primary hardware developer, and with EO's close relationship with AT&T, Rossmann was in a position to make good on his threat to destroy Penpoint — if he was crazy enough to do it. And it was starting to look as if he just might be.

The dispute with EO was only one of our emerging problems. By early spring, it was evident that the sales projections for 1993 were a pipe dream. Virtually all of the sales had been slated for the fourth quarter, which had now slipped into 1994. State Farm accounted for a substantial portion of the shipments, but Norm Vincent had called and reluctantly explained that owing to internal politics, he had to put the auto claims estimating project on hold for a year. As a result, the new $10 million investment from AT&T, which was still being negotiated, would stretch only until September 1993 — assuming that EO accelerated the payment of the final $3 million. As a show of good faith, Bill released the Penpoint code to EO, but was still waiting for the money.

To our surprise, IBM had formed another task force to determine its strategy for pen computers. Our salespeople happily flew to New York to explore this revitalized opportunity, but were disappointed to learn that IBM's new approach was to promote only software to which it had exclusive marketing rights. All the worse, members of the new group were ignorant of Penpoint, and even inquired as to where they might get a demo of their own pen product.

IBM had also managed to destroy what little brand identification it had managed to create around the Thinkpad by co-opting the name for its new line of laptops. This created no end of confusion in the press, which had been following IBM's progress in pen computing only to find the company implicitly declaring a return to keyboard-based machines.

The euphoria of the post-COMDEX publicity for Penpoint had faded considerably, particularly after General Magic had pushed the limits of hype into uncharted territory. It staged a major press announcement in early February with its corporate partners — AT&T, Sony, Motorola, Matsushita, and Phillips — without showing a product and without promising any firm time frames for delivering one. Asked about when General Magic's technology would have a real impact, Marc Porat, the company's visionary CEO, made the strategic mistake of predicting that it could take ten or twenty years. For reporters who lived and died by daily deadlines, this was tantamount to promising salvation in the afterlife. The announcement confused the press about GO's prospects once again, because General Magic proposed to build a new operating system called Magic Cap to host their communications software, and so AT&T appeared, to the casual observer, to be supporting two competing Silicon Valley startups in the personal communicator market.

Not all our problems were external, however. GO's product schedules had slipped as well, raising the ominous possibility that we would need to seek out even more money, at a time when we had nothing to show but a bagful of promises and missed schedules. What was supposed to be a midyear release for the next version of

Penpoint — which was to occupy substantially less memory and be much easier to use — began to slide inexorably toward the dead of winter. Most of the engineers were occupied with simplifying portions of the program, to reduce its size. But in the course of rewriting, everyone was proposing improvements that would tend to increase its size. Penpoint was succumbing to a disease known as feature creep — the irresistible temptation for engineers to load a product down with their favorite special features. Before unnecessary additions hardened around their feet like concrete, the senior technical managers staged meetings called Feature Court, with "Judge Wapner" presiding. Each side would argue its case, and a binding decision was made on the spot. Despite these efforts, the system seemed to grow like a cancer. No sooner were we able to save some space than a new, indispensable requirement would emerge. Bill's confidence in Robert's ability to manage the process began to erode.

To make matters worse, Intel came in to pitch us on porting Penpoint to *its* new chip family, code-named Polar, a hastily planned response to the Hobbit and ARM chips. It emerged that Polar had been designed in conjunction with Microsoft, which wanted a RISC chip for a stripped-down version of Pen Windows that Lloyd Frink was developing, called Winpad. As Robert put it, "The good news is that we have Microsoft on the run. The bad news is that they're running in our direction."

As the depth of these problems became increasingly clear, Bill reacted by stepping up his already breakneck pace. He spun faster and faster, dashing from meeting to airport to conference call to presentation, until he seemed ready to drop. He had personally supervised the launch of Penpoint J and the opening of GO's new office in Japan, often shuttling back and forth across the Pacific twice a week. Randy had warned me that Bill might get seriously ill from all the stress.

But what seemed to finally break Bill's spirit was an incident with Compaq Computer Corporation, a major manufacturer of IBM-compatible personal computers, which had decided to get serious about selling a pen computer. Our sales people worked for months to educate Compaq's executives on the market and our product, bidding against Microsoft for the business. After a formal proposal and presentation to senior Compaq officials, they let us know informally that we were the leading contender. The final step was for the evaluation committee to make their recommendation to the company's CEO. Suddenly our staff's phone calls to Compaq went unreturned. Bill learned shortly afterward that Microsoft, and possibly Bill Gates himself, allegedly countered by proposing to Compaq's CEO a special relationship giving Compaq an edge over other personal computer manufacturers — on the condition that Microsoft get the pen project as well. To protect the company's main business, the CEO agreed.

Knowing he would be in Foster City late in the afternoon on March 1, Bill requested an emergency meeting with Robert, Randy, and me in his office at six. Somehow we all knew he had something important to discuss.

Bill looked like hell. His eyes were bloodshot and his head shook periodically to ward off the desire to fall asleep. "Sit down, gentlemen," he said solemnly. "I've come to a difficult conclusion. Since last August, we've been living a lie. There's no way we're going to pull this through. We're spending over two million a month with no revenue in sight. Only the copy-center people really know what their job is. Everyone's stretched to their limit." As he spoke, he flexed a rubber band between his fingers ever more vigorously, until it snapped. "I want you guys to know that I don't care what happens to me. The important thing is to save the project and the organization — to protect what we've built." While Bill often emphasized his points with strong language, he was doing just the opposite now. His restrained wording carried special weight, especially in this private setting. We weren't sure where he was heading, but it didn't sound good. "Robert, Jerry, I think we're going to have to put the company up for adoption — to find it a good home."

We were momentarily stunned into silence. "You're saying we should sell the company," I offered in clarification.

He couldn't bring himself to use the words, so he just nodded. "Can you guys get comfortable with that?"

Robert and I looked at each other. Robert's eyes said it all. For years, he had been the trouper, ready to go just one more mile. But this time he looked weary in a way he never had before, as though he wanted to say "enough is enough." Perhaps under the wing of a larger, influential company with a long-term perspective and deep pockets, we could rekindle the sense of optimism and purpose that was so essential. We both knew this meant Bill would likely be leaving.

I spoke for both of us. "We won't be a problem. Do what you have to do. We'll back you up all the way."

Bill turned to Randy. "How about you?"

For once, there wasn't a hint of a smile on Randy's face. He spoke slowly and philosophically. "You know what? The problem is simple. We have a great product that doesn't sell. It seems to me that GO is like *Gone with the Wind,* a labor of love with no real possibility of a return. Did you know that Selznick was so obsessed with making the movie that he gave away the first seven years of its profits to another studio just to get Clark Gable? We've built something really special here, and I agree with you guys — the first priority is to preserve the project and our people. I can survive losing my job, but it'd kill me if after all the hard work, brilliant engineering, and money, the product died before it had a real shot in the marketplace."

"We're all in agreement, then," Bill said. "I'll talk to the other estaff. Until I say so, everything's business as usual, OK?" Everyone nodded. "OK, then. We need some runway while I work up some candidate buyers. The most obvious is AT&T, but we should also contact Novell and Lotus, maybe even IBM. Randy, what's the latest on the AT&T investment?"

Randy wasn't expecting to be asked an operational question at a time like this. He had to pause for a moment to get his bearings. "We're just about to close. The problem is the shelf full of interlocking IBM agreements in my office. AT&T discovered that IBM has a right of first refusal on future financings, and now *they* want the same thing, which is impossible. So our lawyers have worked out a formula they call the 'shootout scenario.' Essentially, each of them gets to bid against each other until one of them wins, but the winner has to buy out the loser's interest in the company. It's crazy, but it works. If I can get IBM to go along, we'll get the money in the next week or so. Beyond that — should we need it — is the Taiwan money."

Even though the AT&T financing had yet to close, Dave Atkinson had taken Randy to Taiwan — where AT&T was in the middle of negotiating a major agreement with the government there — to pitch them on investing in GO. We took this as a show of good faith, and it looked as if it was going to pay off. The Taiwan government's representatives expressed a strong interest in buying 10 percent of GO, alongside AT&T's investment. For once, we had excellent prospects for future financing.

In light of this, it seemed peculiar that we were talking about selling off the company. But as I looked around the room, I realized for the first time that money wasn't the problem. It was a loss of faith: Randy's misgivings about the viability of our business; Robert's doubts about our development schedules; Bill's skepticism that we could develop the market momentum required to succeed as an independent company; and my concern that the new management team might not stick it out for however long it took. Despite the expanding bubble of hype about portable computers and mobile communications, our dream seemed as elusive and distant now as it ever had. Besides, Bill was right. When push came to shove, what mattered most was the project itself, the inexplicable imperative that no matter what, the organization had to survive and carry it on. We sat in silence for a while, each of us considering the gravity of the decision we had just made.

"OK, then," Bill said again. "Jerry, you go and see what kind of grass-roots interest there might be at AT&T. Find whoever you can and tee this thing up for me. I'll talk to the outside directors, and Bernie Lacroute. Robert, find a way to keep us on track for an end-of-year product delivery, no matter what. Randy, close on the AT&T money, and start figuring out what pitfalls there might be in our legal agreements for an acquisition."

"There is one," Randy said. "I mean, one that's especially important to us. In the event of an acquisition, the preferred investors get a double dip."

"How's that?" Bill asked.

"It's in the stock purchase agreements. In case of a 'change-of-control transaction' — which this is — preferred shareholders get their original investments back, *then* they convert to common stock and take their pro rata share."

We all knew what that meant. It was one thing for AT&T to feed us a few million at a generous price, but quite another to find a partner willing to swallow the company whole at our current size and burn rate. At the price we could fetch in an acquisition, there might be little or nothing for the common shareholders after the investors got their $50 million back, and the bulk of what was left. As we discussed a possible acquisition, each of us had in the back of our minds that in some fashion we would get a reasonable payoff for all the years of hard work. It might not be a fortune, but it was something for the value we had built.

Robert looked annoyed. "Christ, that really sucks." This was an understatement. At the latest valuation, my shares were worth over $10 million, and with Robert's latest stock options, his were worth around $5 million. "How the hell did that happen?"

I thought back through our negotiations with various investors over the years. Somewhere, in order to close one deal or another, we had acceded to investors' demands for this condition, which is called a "participating preferred." At the time, it had seemed a minor concession, unlikely ever to make a difference if we became successful. But now it meant the difference between a respectable payoff and nothing at all. We were faced with a real-life moral dilemma: either we protected our personal fortunes or gave up our stakes in the service of a dream.

Bill heaved a deep sigh. "I want you guys to know that I don't care if I make a dime out of this."

The three of us swallowed hard, almost in unison, and nodded in agreement. As we walked out, Robert leaned over to me and whispered, "Joyride to hell."

In the next few months I made a grueling series of trips to New Jersey, chasing down whoever would listen to my pitch. Like most native New Yorkers, I had rarely ventured off the New Jersey Turnpike, and so thought of "The Garden State" as a bunch of smelly refineries surrounded by a mass of dull, generic tract houses. To my surprise, behind the ugliness lining the turnpike lay charming hamlets and inns, like some Potemkin village in reverse. South Jersey contained tidy seaside communities alive with salt air; to the north, bucolic colonial manors dotted the countryside, surrounded by flocks of gentle mallards. Even the roadside diners, with their shiny steel siding, had an authentic feel.

It seemed as if AT&T had an office complex unobrusively tucked away in every town. Decades of experience at camouflaging switching stations had spilled over into the siting and design of its corporate facilities — in most cases, you can't find your destination without a guide. Once inside these undistinguished structures, a visitor soon becomes aware of their enormous scale. At the larger complexes, within a sprawling labyrinth of aisles and culs-de-sac, thousands of people labor away in modest offices with obscure designations like "4342K2." With 50,000 workers in New Jersey alone, AT&T employs a significant percentage of the state's workforce. Strictly speaking, AT&T's offices, like its phone networks, aren't really located anywhere — they're everywhere.

For my first few visits, I whistle-stopped my way from town to town, seeking out executives, making my way up the myriad corporate ladders that converged on Bob Kavner's office in corporate headquarters. At every opportunity, I wove a tale about the personal communicator as the successor to the cellular phone, and how the software that drove its user interface was as strategic an asset as the systems that connected the millions of long-distance calls made every day. "Today, dialing a call is like programming in assembly language," I would say. "You punch in a bunch of digits, and it rings a bell at a specific remote location. With the personal communicators of the future, we will address people, not machines, wherever they happen to be. And if they aren't available, you can simply jot them a note."

After working through an endless series of gatekeepers, I finally secured an appointment to meet with a key strategic advisor reporting to Kavner himself. A trim and stylish man with the aspect of an Ivy League professor, he had as his assignment the development of a master plan for the future of what AT&T executives fondly called "the cloud" — the amorphous, ubiquitous web of wires, satellites, relay stations, and cells identified by a familiar audible *bong* followed by the musical rendering of "AT&T," spoken by a sparkling female voice.

On the morning of my appointment, I was awakened at six A.M. by the distorted voice of an evangelist booming from the hotel's clock radio, pitching his own personal formula for getting to heaven. I arrived half an hour early at AT&T corporate headquarters, a magnificent array of stacked, truncated pyramids tucked in the hills of affluent Basking Ridge.

A uniformed guard questioned me at the entrance to an underground garage, constructed as a series of connected caverns, with spiral ramps and gates requiring coded cards for access. "You're going to the executive offices," he exclaimed after finding my name on his list, as though this were an invitation to the Vatican. From the visitors' rotunda on the ground level, I was ushered onto an escalator that rose past a waterfall to an atrium at the top. Here a polite young woman filled out a pass for me to carry.

Her perch looked out at the rear end of an enormous golden sculpture of Wingèd Mercury standing atop a globe. I noticed that he was naked. "How do you like staring at the butt of a Greek god all day?" I asked.

"That's why I took this job," she said, pointing me toward a bank of elevators.

Ascending to the top floor, I was greeted by an attendant at another way station, who directed me to a hostess who managed the fully equipped offices made available to visiting executives. She, in turn, buzzed me through a glass security gate and pointed me toward a magnificently appointed office, paneled in walnut and decorated with abstract art. "I was asked to tell you your meeting may start a little late. Please let me know if you need anything." Here I waited.

The advisor showed up fifteen minutes late, apologized for being tardy, and announced that he had an important conference call in fifteen minutes that should take only a short time. He looked distracted as I launched into my pitch. He soon withdrew to take his call, only to reappear forty-five minutes later. Unfortunately, he was already late for another meeting, so he asked if I wouldn't mind waiting for an hour and a half until it ended. He returned two hours later, and I continued my pitch. Within minutes, he got another conference call. This one lasted over an hour. When he came back, he told me that his wife had been waiting for him downstairs and he had to leave immediately.

I begged him to take a moment to compare our calendars for another open slot. As it happened, he wasn't available again for more than a month. But I found a loophole: he would be in Tokyo ten days hence. Coincidentally, I was scheduled to talk at a convention of Toshiba's marketing partners in Japan that week, and arranged to meet him at his hotel upon my arrival. When he left, I calculated that I had been waiting for him a total of five hours.

Ten days later, after the eleven-hour flight to Tokyo, the two-hour bus trip into the city, and a half-hour cab ride from my hotel to his, I called his room. It was already seven P.M. in Tokyo — two A.M. in San Francisco — and I was starting to hallucinate from sleeplessness. He said he would be down in fifteen minutes. An hour later, he showed up with his wife. We withdrew to a sushi bar, where I launched once more into my pitch while his wife looked on impatiently. By now I had become obsessed with delivering it. I waited for his reaction.

"I hear that IBM has the source code to your software," he said.

"That's right, but they have to keep it compatible with our APIs." I could see that he wasn't clear on what this meant. "I don't think it would be an impediment to anything AT&T might want to do with us."

"Thanks for the input. I'll keep it in mind in our next cycle of planning." He collected his wife's jacket and placed it around her shoulders, then they left.

By mid-April, it was obvious to everyone that something was terribly wrong with the way things were organized. GO, EO, and AT&T Microelectronics had their own separate sales forces calling on the same customers, who were growing increasingly confused about who was responsible for what. GO and EO were each doing software development and courting ISVs to build applications for them. EO could not agree with Microelectronics on a common direction for future Hobbit development. The hidden cost of decentralized management and interorganizational communication was bringing the joint effort to its knees.

To address these issues, Dave Atkinson and Bill Warwick began proposing a concept they called the virtual corporation. Mainly by gaining a larger degree of control over EO and GO, and gathering key managers from all three companies, they hoped to streamline operations. Believing that Kavner supported their approach, they called for meetings to plan the details.

Behind the scenes, Atkinson knew that EO's corporate culture would have to be redirected, so his first priority was to gain a greater degree of control over EO. He began discussions with Bernie Lacroute, EO's chairman, to increase AT&T's equity stake from 24 to 52 percent, on the assumption that this would enable AT&T to direct EO's activities. But owning a majority interest in a startup like EO doesn't necessarily result in control: the company still has to be managed as an independent entity in order to respect the rights of the remaining minority shareholders — something that AT&T was unaccustomed to doing.

The EO board accommodated Atkinson's wishes and worked out a deal in which AT&T put up nearly an additional $40 million to increase its percentage ownership, giving EO a generous valuation of $150 million after the transaction. But instead of investing the full amount directly in the company so the funds could be put to good use pursuing the project, about half of it went right into the pockets of EO's common and preferred shareholders. The reason for this unusual arrangement was that it cost AT&T less to reach its goal by purchasing existing investors' shares than if it had purchased enough newly issued shares to dilute all other stockholders down to a 48 percent stake.

Kleiner Perkins agreed to sell half of the EO stock it had purchased barely two years earlier, for two and a half times as much as it originally paid for the whole lot. KP netted about $7 million — as much cash as it had invested in GO and EO combined.

The deal proved to be a bonanza for many of the individual stockholders, despite the fact that the company was far from turning a profit. Hermann Hauser of Active Book sold one third of his stock to AT&T, netting him several million dollars. The earliest GO employees, who had stoically suffered through five years of perpetual development, were aghast to learn that some of their colleagues who went over to EO had pocketed small fortunes while they were still toiling away. As one GO software engineer observed, "AT&T is the *Beverly Hillbillies* of venture capital."

What AT&T didn't realize was that it had just wasted its money, since EO maintained business as usual. But to those responsible for budgetary planning, it looked as though AT&T had just invested an additional $40 million in its personal-communicator initiative. So when Kavner's staff began thinking about taking a similar position in GO, he wouldn't hear of it.

Under growing pressure to focus GO solely on AT&T's goals, Bill Campbell was increasingly concerned that Kavner might not be sold on the virtual-corporation concept. To allay Bill's fears, Atkinson and Warwick arranged to have him meet with Kavner while he was on the West Coast for an EO board meeting. What Bill didn't know was that the EO board had just rejected the contractual revisions and $3 million payment that GO and EO had recently agreed to — something he assumed Rossmann had the power to decide on his own.

Warwick ushered Bill into the main conference room at EO, where he took a seat at one end of the large conference table, opposite the end where Kavner was holding court. Warwick began the meeting with a detailed explanation of the virtual-corporation plan. Bill focused his attention on Kavner, who was squirming in his seat, plainly skeptical about the idea. Unaware of Kavner's discomfort, Warwick turned the meeting over to Bill.

Bill aimed his remarks directly at Kavner, as though no one else were present. "Let me tell you what we're trying to accomplish. We have three organizations represented here. In many cases we're working together, but in many we're working at cross purposes. We really believe that the AT&T umbrella ought to be over all three of these. They don't have to be one organization, but to the outside world it has to look like we're thinking and acting in unison, and that AT&T is fully behind it."

Kavner looked suspicious, and quickly served up his concern. "I don't think AT&T should own an operating system company. Why should we? What we want is strong, independent partners who can stand on their own and make money."

Bill was getting annoyed. He had been led to believe that Kavner was already on board, but instead Kavner seemed to be accusing him of a hidden agenda. "Whether you want to own an operating system or not is up to you," Bill shot back. "I'd like to make money on my own, take the company and go public. But I'm being urged by your people to devote all my effort to the Hobbit chip and to AT&T,

flying to Japan eight million times and New Jersey eight million and one. But I'll tell you why you *should* want to own an operating system. There are two key parts to the intellectual property, the processor and the OS, and right now you only own one."

As the exchange grew more heated, heads silently followed the action from one end of the table to the other, as though watching a tennis match. Kavner got to the point. "It looks to me as though GO is running out of money. You're trying to get several million dollars out of EO for some minor contract violation!"

Now Bill believed that somehow, somewhere, he had been set up. Unaware that Rossmann had brought the contract issue to the board, he thought it had been completely settled before he agreed to meet with Kavner, despite the fact that he hadn't seen any money yet. Nevertheless, he didn't think it was appropriate for Kavner to get involved in a matter of this sort. "That's a business issue between GO and EO, and you should stay out of it," Bill said. "I can see this meeting is a waste of time." He abruptly ended the discussion and walked out of the room.

Atkinson was close behind, and caught up with Bill just outside the building. "I didn't know anything about this!" he protested.

Bill laid into him anyway. "I don't know what cave you've been in, Dave, but I could smell this from three thousand miles away. Kavner's nowhere on this. Nowhere." Bill got into his car and drove away.

For days after the meeting with Kavner, Bill refused to talk to Lacroute or Rossmann, despite several phone messages from both of them. Things seemed to be at a standstill until Lacroute convinced Kavner to call Bill personally. They had a tense but cordial chat, and Bill permitted discussions between GO, EO, and AT&T to resume. Since it was clear that Kavner wasn't going to be a sugar daddy for GO, and we still hadn't got the $3 million due from EO, we continued looking for alternative financing, such as the proposed investment from Taiwan. But on Tuesday, June 22, a bombshell hit.

Bernie Lacroute called to deliver the bad news. "Bill, Kavner's gotten serious about ditching Penpoint and going with Newton."

The mere suggestion seemed wacky, given Kavner's public statements of long-term support. "What the hell's going on?" Bill said. "Just last week he advised Matsushita to get behind Penpoint!"

"I don't know," Lacroute said. "But he's asked his team — and us — to go down to Cupertino and check it out."

Bill considered the consequences. "If that happens, we might as well close the doors. And it's no picnic for you guys either. What on earth could make him even think of such a thing?"

"I'm not sure. But we have to get a proposal in front of him right away to sort out the GO–EO situation or I'm afraid we'll lose everything. I've got a dinner with him next Monday. He's willing to consider a plan to merge the two companies . . ."

Bill didn't even bother to put down the receiver before making the next call. He had moles all over Apple, and by the end of the day he had assembled the whole story. Behind the scenes, things were looking pretty sour in Cupertino. Sculley was coming under fire for hobnobbing with politicians when he should have been paying more attention to business. Sculley reportedly met with Bob Allen, the CEO of AT&T and Bob Kavner's boss, to discuss the possibility that the phone giant acquire Apple outright. But Allen's staff had decided that Apple's prospects weren't rosy enough, so they began considering a less ambitious goal: a joint venture between Apple and AT&T in the area of consumer products. Sculley flew cross-country with Kavner on one of AT&T's corporate jets to explore the idea of combining their efforts in this area, along the lines of the deals he had cut with IBM. Disappointed with EO's and GO's product delays, Kavner apparently wondered whether he should throw EO into the pot and switch AT&T from Penpoint to Newton.

Bill called an emergency board meeting for six o'clock Tuesday evening to consider the options. John Doerr was ill with the flu, and so attended by phone. Without being able to read Bill's body language, John was less restrained than usual in proposing new partnerships, but the converse was also true: Bill was freer to express his reactions with animated gestures. After a while, the meeting looked like a game of charades.

Although no one liked the prospect of merging with EO, the board worked out a message for Lacroute to take back to Kavner:

Yes, we were willing to talk about a merger, but even better would be a joint investment on the part of AT&T along with another partner, such as Novell or even IBM, or at least AT&T's support in finding another partner.

Given Lacroute's personal stake in EO, Bill wondered whether he could fairly represent GO's interests to Kavner. Bill put the question to the outside members of the board. John and Vinod Khosla stood behind Lacroute without reservation. David Liddle concurred. "If Bernie agrees to put on his KP hat, he'll wear it with Kavner."

John, over the phone, attempted a bit of levity. "If we do wind up merging the two companies, let's call it GEO and not EGO." At that moment, David opened his Apple Powerbook, which was sitting next to the speaker phone, and it said "*David's got a headache*" in a Monty Python–esque voice. It seemed he had replaced the usual startup beep with this customized announcement.

"Who the hell was that?" Doerr asked.

"David's Powerbook doesn't like your joke," Bill said.

After the meeting, Bill called Lacroute at home, despite the late hour, to deliver the message. Lacroute sounded exasperated. "You don't get it, you just don't get it. Kavner doesn't want options, he wants us to merge the two companies. It's that or he's going with Apple!"

"OK, OK, I hear you," said Bill. "But Bernie, we both know there's no such thing as a merger in Silicon Valley. One company eats the other, and only one survives. And under these circumstances, it's important that we don't look like we're fighting among ourselves."

Lacroute mistakenly took Bill's comments as self-serving — that he was looking out for his own job. He decided to sound Bill out on running the merged companies. "Bill, would you consider being the CEO?"

Bill took a deep breath. "No way. Rossmann can handle the job." Bernie was astonished. "Why not?"

"AT&T has already committed to making EO their flagship company in Silicon Valley. And besides" — Bill let the receiver drop around his chin — "life's too short to work for AT&T."

The following Monday, Bill and I flew to New Jersey with several other GO people to make a presentation the next day to AT&T's global business systems division. The flight arrived late at Newark Airport, delivering us into the steamy summer heat. A heavy shower had just ended, and the sticky pavement seemed to glow black under the harsh bluish light of the streetlamps. Bill cursed loudly when our rental car wasn't ready, forcing him to wait in a long line of sweating salesmen and charter-flight tourists.

Bill drove west like a maniac to our hotel near Basking Ridge. For some unfathomable reason, the New Jersey highway authorities were in the process of relabeling the exits on Route 237, with signs like "Exit 35, formerly Exit 33." Unclear which turnoff we wanted, Bill took a chance, so we wound up backtracking through what appeared in the dark to be endless pastures. After dropping us at our hotel, Bill went careering off to the Short Hills Hilton, where he was to meet Lacroute after his dinner with Kavner.

The meeting was considerably more formal than Bill had imagined. Not only were Lacroute and Kavner there, but also Rossmann, Atkinson, and representatives of a number of other AT&T divisions. Bill cooled his heels in the lobby of the hotel, pretending to read while the fate of our company was being decided in a rented conference room above. Each time the elevator bell rang, he would quickly lower his magazine, raising it back up slowly when it turned out to be a false alarm. Finally, after midnight, Lacroute, Atkinson, and Rossmann emerged wearing broad smiles. "Let's go sit in the bar," Bill suggested.

Dave Atkinson did most of the talking. Kavner had looked disinclined to stick with EO and GO at first, but when he had gone around the room polling his lieutenants, only one was in favor of working with Apple instead. Although Kavner had liked the idea of standing shoulder to shoulder with John Sculley at a public announcement, he knew that it was imprudent to overrule his advisors. In the end, he told them to proceed with the GO–EO merger, but to keep the door open to Apple. It was a wise move: soon afterward, Sculley left Apple.

"You did a good job lining up support in the ranks," Atkinson said to Bill.

But that wasn't all. There were a number of conditions, which

Atkinson relayed to Bill as gently and sympathetically as possible. Kavner wasn't willing to put up any more money, and when the merger was done, he still wanted voting control of the combined companies.

"What about the Taiwan investment?" Bill asked.

"We're going to ask them to redirect that money into EO."

"What does Warwick think about this?" Bill couldn't believe that his friend and staunch supporter was in favor of this plan.

"It doesn't matter," Atkinson said. "He's just been reassigned to China."

Bill found the nearest pay phone and called me in my hotel room with the news. I was livid. "So instead of AT&T paying us for the company, we're supposed to hand it over to them in return for funny money — restricted EO stock. Then we're supposed to figure out a way that they get to vote that stock for us, and pray that someday they allow us to cash out."

A recorded voice interrupted me — "*Insert twenty-five cents for the next three minutes, please*" — but Bill didn't have another quarter.

"I don't get it," I said. "We could have gotten by without their money. All they had to do was stand firmly behind us."

"Some fuckin' life, huh?" Bill said. Then the line went dead.

14

The Showdown

WHEN I WAS FIVE, my parents got a portable television set for my bedroom. At first I was afraid to get dressed in front of it, for fear that the people on the screen could see me, just as I could see them. I knew that the gray, blurry pictures were merely signals magically sent through the air, but I didn't understand which qualities of living human beings carried over to the new machine. Like the rest of my generation, I was captivated by this limitless window on the world, and what I saw seemed as real to me as my own two hands.

Through the power of television, I knew that lions were large, proud creatures that roared and lived in Africa. But one day my father took me to see the big cats at the Bronx Zoo. I could barely relate the coarse, smelly creatures in front of me — lying docile and panting in their cages while flies buzzed around their heads — to what I had witnessed from the comfort of my bedroom. The TV images just didn't capture the real experience.

Sitting in the briefing rooms of AT&T thirty-five years later, watching staffers project spreadsheets, graphs, and slides from their portable computers onto the large screen monitors, I realized that computers mislead managers just as television misleads kids. Fed a steady diet of numbers and charts from the comfort of their conference room chairs, senior executives experience only a desiccated version of the powerful forces that shape and grow their organizations. Mistaking these two-dimensional reflections for reality, they shadowbox their way through complex decisions, unwittingly jostled in one direction or another by self-interested emissaries, who can spin a tale of threat and opportunity as skillfully as any Hollywood screenwriter.

In these rooms, individuals are stripped of their unique skills and reduced to chits, then shifted from one column to another. Complex working relationships, knitted together over time like trees intertwining their limbs, become statistical learning curves. Loyalty and trust, painstakingly earned through years of delivering useful products and serving customers' needs, are measured as the difference between market capitalization and book value. Real human wealth, in the form of security, freedom, productivity, and knowledge, is scarcely captured by unexercised stock options. There are no line items that gauge the real engines of prosperity: vision, passion, and commitment. The plain truth is you cannot suck reality from the hypnotic glow of a vacuum tube.

While Bill and Randy began planning the merger, my job was to keep GO's visibility as high as possible among the powers at AT&T. I was lucky to be invited to present some of our most recent work to Bob Kavner at an early morning briefing, barely two weeks after his powwow in Short Hills. Flying to New Jersey for a planning session, I spent an entire day in a room full of nervous staffers, who reviewed my slides word by word, carefully coordinating my presentation with theirs.

"When you do your demo," one staffer suggested, "it's a good idea to focus your benefits around increasing long-distance traffic. That's the real mother lode around here."

"Got it."

After ten hours of review for a fifteen-minute demo, the next morning I walked into the conference room where Bob Kavner and several other key executives sipped coffee and chatted about the news: Kavner had just been promoted to CEO of the multimedia products and services group. One of GO's sharpest salespeople, Danny Shader, was accompanying me on the trip. As unobtrusively as possible, he went around the room collecting business cards.

"What'd you do that for?" I whispered. "You pretty much know who everyone is."

"Titles are deceptive, but you can tell who really matters by the paper." He showed me the cards. "The ones with the real power are printed on this bone-colored stock. The others have white cards with blue logos."

I sat down next to Kavner and made my pitch. "Most people think that you have to have a computer with a keyboard to access a database. But there are thousands of VTR [voice tone response] systems in operation today. You know, 'Press 1 to hear your balance, press 2 to hear your last five debits.' In fact, this is probably the dominant form of consumer data access today. Working with your people, we've prototyped a new, richer telephone interface under Penpoint — the successor to today's twelve-key pad — which integrates voice and data for the phones of the future."

Kavner watched the screen closely throughout my demo, nodding periodically. But whenever he had a question, he turned to his own staff for an answer. I felt as though I were just another audiovisual device delivering the latest briefing in a more personal form.

I groused to Danny as we packed up our gear. "I felt like a trained monkey."

"You looked like a trained monkey."

"What a waste of time. We fly all the way out here and squander a whole day in a planning session, then get to entertain Kavner for half an hour before his real day gets under way."

"You don't get it," Danny said. "This goes on all the time. It's just how things get done in a big company."

I hadn't thought about it before, but I had never actually worked for a big company. And based on what I had seen, I wasn't so sure I wanted to.

The next estaff meeting was superseded by a joint executive staff meeting of GO and EO, to start planning the merger. As usual, most of us were in the men's room, unloading before heading over to the conference room in another part of the building. Robert and I occupied the two urinals, one of which was much lower than the other, no doubt to satisfy some building code. Bill and Randy waited their turns.

For some reason, Bill decided it was a good time for a team huddle. "Listen guys, I know there's been a lot of bad blood between us and the EO folks, but it's up to us to help make this thing successful. Our people are going to be counting on us to do the right thing. So remember, this meeting isn't a contest to see whose dick is bigger."

Robert leaned over toward the privacy partition. "I see you're using the kiddie urinal," he said.

"This one's for the guys with the big dicks," I responded.

"You're right — this water sure is cold," he shot back.

Once the EO folks arrived, led by Alain Rossmann, the first order of business was to compare our two organization charts for duplicated functions, which of course were everywhere. "I want to keep our run rate under thirty million after the merger," Rossmann said. "We've got about a hundred fifty full-time heads, and you've got over two hundred. We need to target no more than two fifty after the merger." As we reviewed each area, the managers responsible deliberated over the strengths and weaknesses of their key people. Gradually a picture of the combined organizations emerged, and it became clear that GO needed to lay off eighty to one hundred people. The plan was for each manager to produce a list of who was to go and who was to stay.

After the meeting, Bill and Randy met with Rossmann to discuss other issues. Calmness prevailed until they got to the matter of how to value the two companies in the merger. Bill started with a simple offer. "The last post-money valuations of our organizations were about the same, so I think the fair thing to do is just make it a fifty-fifty deal."

Predictably, Rossmann had a different perspective. "Your last valuation was based on a small investment — ten million from AT&T — at a high price. Ours was a real transaction. I'll give you one third, tops."

"Maybe we should figure out first whether you're buying or I'm selling."

"And by the way," Rossmann said, "you have to eat the IBM loans."

Bill lost his cool. "This is bullshit. I'm going to talk to Atkinson. You knew about the IBM loans all the time. I'm not going to take less than forty percent. And the loans are *your* problem."

It was beginning to look as if the two of them might come to blows. Randy jumped in to cool things down. "We could scream at each other about this all day. Why don't we just get an investment banker who we both trust and ask him to figure out the relative valuations."

This seemed sensible enough. Within a week, both sides had agreed on Ruthann Quindlen, an informal and fair-minded young banker who had previously spent ten years with the firm of Alex Brown and who had plenty of experience pricing computer companies for the public market. She was already familiar with the situation, having recently married one of our board members, David Liddle.

◫

A few days later, I had a chance to meet with Mitchell Kapor and bring him up to date. He and his family, on vacation in San Francisco, were staying with John Doerr. Layne and I stopped by with some Indian takeout food.

Mitchell's two small children, a boy and a girl, had just learned to tell riddles.

"When is a dog not a dog?" the boy announced to the crowd.

No one answered.

"When it's a *hot dog!*" he said. The two of them giggled wildly.

"Want to hear it again?" he shouted. "When is a dog not a dog?"

The girl answered in a high-pitched squeal. "Den it's a *ho-do!*" They both laughed again, no less enthusiastically.

Mitchell and I went into another room while the children started a third round. "They haven't quite got it yet," he explained, "but it's amazing how much I love watching them learn."

I could see that something had changed in him. He seemed less intense, more at peace with himself. "Where did you park the jet?"

He sighed. "Sold it. I'm also selling our vacation house in Hawaii."

"Too bad, I never got to mooch a vacation there," I said. "How come?"

"We just decided to get our priorities straight. All this stuff — our material possessions — were weighing us down. They owned us instead of the other way around."

"Sounds great. What convinced you to do it?"

"Nothing, really. We just started to realize that life is about people, not things. We were wrapping ourselves in stuff to prop up our egos. I guess we just didn't want our kids to grow up thinking that what you have is who you are."

I explained to him about the merger. I thought he might be concerned about the million dollars he stood to lose.

"Just hand it over to those turkeys and say good riddance," he said. "They're just going to yutz it up anyway. The war's over, and you may not believe it, but you won. Now it's a deep-pockets, big-company thing. The important thing now is to do what's best for your people and the project."

"What do you mean, we won? We've spent nearly seventy-five million dollars, the product didn't sell, and AT&T is stealing the company out from under us."

He waved his hand as though pushing my comment aside. "Have you thought about how many jobs you created? how many people you inspired? how many people believe in your vision of pen computing? These things will last a lot longer than a bunch of stock certificates."

As the last orange rays of the sun slipped out through the Golden Gate Bridge, I walked over and gave Mitchell a hug.

Within a week, Ruthann Quindlen was ready to present her results. "You're not going to like this," she said as she passed some papers around the small conference table in Randy's office. "I looked at the market capitalization for some PC software companies of a similar size, then subtracted your outstanding loans. It came to fifty-two million. Then I took EO's last valuation, subtracted a twenty-five percent premium that AT&T should have paid to get control, and removed their obligations as well. That's one hundred and two million."

Bill couldn't hide his disappointment as he wrote the key figures on Randy's whiteboard. "Two to one," he said. "That's essentially what Rossmann is offering — one third, except for the loans. It's tough to swallow, but we agreed to honor your assessment."

We thanked her, and she left to deliver the same message to EO.

Bill erased the board. "According to Debbie B, if we don't get the three million EO owes us, we'll have to shut the doors in two weeks — by August 13 at the latest — if we want to pay everyone their accumulated vacation time and some reasonable severance." Debbie

Biondolillo was GO's vice president of human resources. An old friend of Bill's, she had agreed to help us out by working part time after leaving her job at Apple.

"There's no way Rossmann's going to send over that money," Randy said. "Even I wouldn't do it in his shoes. He should just starve us out." Randy was right, but for the wrong reasons. We still didn't know that the EO board had refused to approve the payment.

Randy started working the keys on his pocket calculator. "There's a small silver lining to Ruthann's numbers. After the preferred investors take their cut, there's still about five-point-two mil left over for the common. That works out to somewhere between fifty and seventy-five cents a share, depending on who exercises their stock options."

"It's a far cry from the three-sixty AT&T paid in the last round," I said. "Our people aren't going to be real happy about that."

Randy swiveled his chair and put his feet on the desk in resignation. "It's better than nothing!"

Bill's voice took on a dark tone. "It's all bullshit anyway. I'll bet each of you one thousand dollars that you never see a nickel for your stock."

Neither Randy nor I took the bet.

The next morning, Randy and I went over to Kleiner Perkins for a preliminary working session with GO's and EO's attorneys on the legal structure of the merger. They recommended a "reverse triangular merger," keeping the legal shell of GO alive but transferring ownership to EO, which had certain tax and liability benefits.

"Kind of like Norman Bates in *Psycho*," Randy observed. "We keep the mummified body in a rocking chair in the basement."

When the meeting ended, I went back to GO. Bill joined Randy and John at KP for a negotiating session with Atkinson, Rossmann, and Dave Kinser, EO's chief financial officer. Kinser, a quiet, polite, youthful-looking man with the short blond hair of a swimmer, had previously worked with Bill and Randy at Claris, so he knew them both quite well.

Bill expected to ratify Quindlen's conclusions as is and get on with it. But Rossmann had other plans. "You know, I spent the weekend playing with Newton. It's good, it's really good. I'm not so sure we should do this deal at all. Kavner's still thinking about the thing with Apple." Rossmann knew that no one at GO had had an opportunity to examine a Newton yet, and he was using this fact to keep up the pressure.

Bill's response was direct and swift. "Don't be an ass."

Rossmann continued, unfazed. "I think Ruthann's analysis is a good starting point, but now I feel that your value is about half of what she recommends — four to one."

"Forget it," Bill said. "It's two to one or nothing. Anything less than that and we'll shut the company down."

"You can't do that," said Rossmann. "It makes no sense. Something is better than nothing. The investors won't let you just throw everything away." He gestured with his hand as though disposing of a used tissue.

"Then the investors can fire me and do the deal themselves. See how many of our people stay when they find out that the management team resigned rather than agree to this."

Atkinson was turning white. He had expected to deliver this deal to Kavner, and now it was starting to look messy. John Doerr was equally nervous. He suggested that the two sides adjourn to separate rooms.

Once back in his own office, John became panicky. "What's the matter with you, Bill? This thing could fall apart any minute. Kavner could change his mind at any time!"

"No way, John," Bill said. "My sources at Apple tell me that the AT&T deal is already dead."

John was pacing around like a ferret. Just then, another of the KP partners stuck his head in to see what the ruckus was about. "I've got a suggestion," he said. "It's not unheard-of for the preferred stockholders to give a bit extra to the common, to secure their vote to get the deal done." In fact, he had discussed this possibility with John in some detail earlier in the day.

Randy saw his opening. He might not have been in a position to negotiate with EO, but he could surely negotiate with KP. "John, if you're so damn anxious to close this deal at four to one, all you have to do is make us whole."

John was livid. Unbeknownst to Randy, he was fully prepared to recommend to the investors that they cut the common stockholders in on the deal, but he bristled at the prospect that it might now look as though he were caving in to Randy's pressure. "Don't you dare try to negotiate with me!"

Randy knew that in John's agitated state, the best thing to do was to stare him down. He walked right over and put his face within an inch of John's. They both stood there, with eyes wide, for several seconds. Then John decided to put common sense ahead of ego. He turned his back to Randy and said, "The average percentage of common in our IPOs is 18 percent. That should give you about the same amount as you're entitled to get at two to one. Now, I've got work to do. Let's talk again later."

Randy seemed satisfied with this and sat back down.

I later learned from Dave Kinser that he, Rossmann, and Atkinson were in the other room while this drama unfolded, evaluating the chances that Campbell might actually shut the company down. Rossmann thought it was out of the question. Kinser believed otherwise, and told a story about the time he and Bill went to the Super Bowl in New Orleans. Bill had two extra tickets that he planned on scalping at the gate. But when they got there, it turned out that prices were dropping by the minute. As they walked along the line of buyers, Bill refused all offers because they were below what he thought was fair — the face value of the tickets. Two kids with no money were standing at the end of the line, hoping somehow to get in. Bill ended up giving them his extra tickets for free rather than take less than what he felt they were worth.

Randy and Bill returned to the conference room, where the others were already seated. "It's two to one or nothing," Bill announced, still in an agitated state. He started gathering up his papers. "Take it or leave it. Call us with your final decision."

Bill started to calm down once he and Randy were in his car. "Why do you think John seemed more concerned with getting the deal done than with getting the best price?"

Randy thought for a minute, then it struck him. "I believe KP owns about the same percentage of both GO and EO. That makes them indifferent to the price."

"How's that?" Bill asked.

"Think about it. Right now I think they own about nine percent of GO and EO. They get paid for their GO stock with EO stock. That means when the deal is made and the companies are merged,

they still own the same thing, about nine percent of everything."
(Randy's logic was correct, but he didn't realize that KP had sold
half of its nine percent share of EO to AT&T in July.)

Bill was worried. "This is a bad scene. John's office is right next
to Lacroute's, and God knows what he's going to say to him. He
wouldn't knowingly screw us, but he just might inadvertently un-
dermine our negotiating position. We've got to find a polite way to
get him out of the loop."

◻

"Really, it's OK, you can go ahead and open it up."

The maintenance man at the Foster City Holiday Inn stared at
his supervisor in disbelief. He had fully expected the night manager
to call the cops, not yield to some lunatic's demand. He had seen
some pretty bizarre stuff while working nights at the hotel, but
never before had a guest insisted that he open the Sports Bar at
seven o'clock on a Thursday morning. The guy didn't look like a
boozer or anything — he was neatly dressed in slacks and a sport
jacket. But if he was that desperate for an eye opener, why not just
raid the minibar in the privacy of his own room?

But Mike Homer had other things in mind. Jumping behind the
bar, he located the equipment that controlled the rooftop satellite
dish. When he flipped on the power, the four giant projection screens
flickered to life. Within minutes, he had reset the transponder code
to the number written on a small piece of paper that he fished from
his breast pocket.

By now, I had joined him behind the bar. "Got it," Mike said,
pointing to one of the screens. A test pattern showed a countdown
to a closed-circuit transmission that would start in forty-three minutes.

"Are you sure this is legal?" I asked. He paused briefly, then
shrugged.

Slowly the room started to fill with product managers, programmers, salespeople, even administrative assistants. Since word got
around that Mike had intercepted the downlink code Apple Computer
was using to broadcast its Newton sales training to regional offices,
and that he had persuaded the hotel staff to open up the bar, many
people from GO decided to show up. By eight A.M., when the broad-

cast was scheduled to begin, fully half the company was packed into the room, which still smelled of stale beer from the last Giants game.

"There's a two-cup minimum on the coffee," Mike joked. "And no, the bar's not open." He stood up on the small bandstand to brief everyone about what to expect. "Apple normally previews important new products to their field people a few days before the public announcement, so they can be ready to answer questions. What you're about to see is a live broadcast from their studios in Cupertino."

With little fanfare, a brief introductory video of the Newton spilled into the room, covertly siphoned off from the flood of electrons bouncing off a geosynchronous satellite 22,500 miles above the earth. Then Peter Hirshberg, the director of enterprise marketing of Apple, appeared holding a Newton, and launched into a demonstration. Notebooks flew open around the room when he began a detailed review of the processor, memory, battery, and input-output subsystems of the unit, which lay dissected on a table in front of him. He gave tips on how to get the best handwriting-recognition results, explained how to identify appropriate customers, and discussed some of the key uses that Apple envisioned for the new product. Next, an engineer talked about the variety of applications currently under development, both inside and outside Apple. The show finished up with a panel of engineers and marketing executives who answered questions phoned in from the audience, and a warning not to expect the product to ship in any quantity for some time.

The mood in the bar was one of relief. Newton was good, as expected, but it fell far short of the all-powerful image created by Apple's formidable hype machine. Even in this short demo, it was evident that the marketing managers' lack of real customer experience had caused them to repeat some of the mistakes of the earliest version of Penpoint. The enemy had shown its face, and it was human after all.

Mike gave a quick summary of the key points, then exhorted the group to get back to work. Instead, people spontaneously started pounding the tables, chanting "Pen-point! Pen-point! Pen-point!" Through the neon signs hanging in the window facing the hotel lobby, I noticed a group of Japanese businessmen getting off the elevator. They shook their heads in dismay at the rowdy Americans packed into a bar having a drink before going to work.

Following the abortive negotiations at Kleiner Perkins, a waiting game began. Bill speculated that the other side might be working on a new proposal, but Randy thought it was mainly an attempt to run us out of cash. Day after day, Randy, Bill, Robert, and I would stand vigil by our phones, hoping for a scrap of good news, a ray of hope, a brainstorm that would break the impasse. But none came. For years we had to run faster and faster just to keep up, and now it seemed that the earth had suddenly slowed to a crawl.

Overtaken by boredom, we would look for excuses to visit one another's offices. First Randy would come over to mine, sit on my threadbare futon, and toss around an inflatable world globe I had received as a promotional gift. Then we would go see if Robert had made any plans for lunch. For the first time ever, Bill began whittling down the piles of newspapers and magazines that grew unchecked by his desk.

Despite our best efforts to remain upbeat, paralysis seeped through the company like a corporeal disease. The nightly software builds took longer and longer, until they stretched well into the next day. My office, always the last refuge for customer and ISV problems, fielded an increasing number of complaints of lost messages and slow responses, as though the staff were going senile. Interoffice mail deliveries — that frayed, dog-eared pile of reusable yellow envelopes riddled with holes — slowed to a snail's pace. The flow of messages on the electronic bulletin boards, always a good measure of the company's mood, weakened like a fading pulse.

Finally Bill called me into his office. Hoping for some breakthrough, I nearly bowled over a group of product testers standing in the hallway who were reading the OSHA notices tacked up next to the vending machines. Bill gave me a moment to catch my breath. "Jerry, I've got a big problem I need your help with."

"Sure, anything."

"Months ago, I agreed to speak at Advanced Micro Devices' annual sales meeting in Hawaii this week. This is a really big deal for them, and I just can't bag it at the last minute like this. You can get a flight out Wednesday afternoon and be back here by Thursday evening. Nothing's likely to happen while you're gone."

I was suddenly overcome with an irrational dread that the calm would break the moment I left. "I couldn't do that," I replied.

"Don't be silly. After all, how much can things change in a few hours?"

Given that nothing had changed for the past two weeks, he had a point. "You'll call me right away if anything happens?"

"Absolutely."

I couldn't imagine why my premonition was so strong, until my mother happened to call my office on Wednesday, just as I was supposed to leave. "You better get going," she said. "You don't want to miss the flight the way you did the evening your father died." She had nailed it exactly. I was living a replay of that horrible night.

When I arrived in Honolulu, the air was filled with the smell of sweet tropical flowers. Advanced Micro Devices had sent a limousine to take me straight to the rehearsals. The company was staging a week-long extravaganza — a pep rally for their sales force. The next afternoon, after I had given my speech and just before heading back to the airport, I called Robert from a pay phone near the bellman's stand in my hotel, to see what was up.

"You mean you haven't heard?" he said. "Surely everybody's been trying to reach you."

"What? What?"

"Just after you left, Bill got a call from Atkinson. They went to dinner with Rossmann and he presented a new proposal."

"What was it?"

"Completely bogus. It was mainly a bunch of warrants instead of stock. They're offering two million shares of common, and warrants to buy twenty million more shares at two dollars a share, exercisable within three years."

"Warrants! At that price, they're virtually worthless!"

"Randy's analysis puts it at thirty to one. He thinks the proposal is worth about three percent of the merged companies. Rossmann wants us to *give* GO to him. Randy was right: this whole time they've been waiting for us to run out of cash."

My limo driver had already loaded my bags and was standing by the curb, trying to look patient. "Look," I said. "I've got to run. Tell Bill I'll call him from the airport."

As soon as I was checked in at the gate, I phoned in. Both the estaff and the senior managers were meeting together. I asked the receptionist to ring the phone in the conference room. Randy picked up the call. "I'll put you on the speaker phone," he said.

"Jerry, are you there?" It was Bill's voice.

"Yeah, go ahead."

"I just briefed the staff about the negotiations with EO. Randy, Robert, and I are about to head over to KP for the final discussion. We're also trying to finish up the cut list. The human resources and accounting staffs are going to stay here as long as it takes to prepare the paperwork for the people we have to lay off. This way, we can give everyone their last checks tomorrow. The only question is whether we will cut eighty people or close down completely. We're still a few heads short."

"Can I say something?" I said.

"Go ahead."

As I spoke, the scene in front of my eyes seemed to dim. It was as if my soul were flowing through the phone line to the room I could picture so clearly thousands of miles away. "I just want each and every one of you to know what an honor it's been for me to work with you, and how painful it is for me not to be there to deliver the news in person along with Bill." I took a deep breath. "Debbie B, are you there?"

"Yes, I am."

"How many heads are we short on the cut list?"

"Two, maybe three."

I knew what we had to do, and even from that distance, I knew that Robert knew as well. Despite our close spiritual connection with the company, we were now expendable. "Robert?"

"I'm here."

"I think our time has come."

"Yep. Let's do it."

"Debbie, please put Robert and me on the cut list. No special treatment. Accrued vacation plus two weeks."

There was a long silence on the other end of the line. Then the head of documentation spoke up, one of the earliest hires. "Jerry, this is John Zussman."

"Hi, John."

"I'm on the cut list too. This situation really sucks, but I just want you to know that GO was a great company and an incredible ride, and I'm grateful to you and Robert for the chance to take it."

A number of people around the table murmured their agreement, then Danny Shader yelled out, "Call me when you start your next company!"

"Yeah, yeah!" A cheer went up around the table, followed by a long round of applause.

I looked up, only to notice that the door to the jetway was closed. "Oh, shit! I've got to run. I'll call you back from the airplane."

Bill said into the speaker phone, "Call us in half an hour over at KP and we'll conference you in."

I dashed to the gate and pounded on the door. Luckily, an attendant was just inside. "Hold it, one more," he yelled to his counterpart at the other end of the jetway. "Can't cut it much closer than that," he said as I ducked under the half-closed hatch.

Once settled into my seat, I looked around for a phone. None on the wall in the front of the cabin. None in the seat back in front of me. None in the armrest. "Where the hell's the phone?" I shouted to the flight attendant.

The attendant walked over to my seat. "We just bought this equipment from Pan Am. I think it's the only 747 in the fleet that doesn't have Airphone service."

I was hosed. Trapped in a slender metal tube hurling across the Pacific Ocean at six hundred miles an hour, I was going absolutely nowhere for the next five hours. Meanwhile, the fate of the company, the jobs of my friends, and possibly the future of the pen-computing industry were being decided in a conference room thousands of miles away.

▣

When Randy, Bill, and Robert arrived at Kleiner Perkins around seven forty-five P.M., they went immediately to John Doerr's office, where he arranged a conference call with David Liddle, who was at home, and Vinod Khosla, who was vacationing with his family in

India. Rossmann and Atkinson were camped out in the cavernous partners' conference room, dug in for a long night of back and forth. Bernie Lacroute, who had a dinner meeting, wasn't expected back until after ten.

Bill began by reviewing what he had told his people back at the office. "This afternoon, I instructed the staff to prepare two alternatives: one where we would be acquired by EO, focus on the Hobbit, and cut down from two hundred to one hundred twenty people, and the other was that we would close the doors and pay all two hundred people severance." His blunt statement set the tone for the next two hours of discussion.

Next, John reviewed the proposal that he and Lacroute had worked up on his whiteboard earlier in the day. John had managed to make some headway over the previous night's term sheet: the exercise price of the warrants had dropped from two dollars to one, which was a substantial improvement. But as far as Bill was concerned, it was still too low even to consider. "No fucking warrants, John. Do you understand me?"

David was more politic. "Warrants are a stab in the heart of the deal that Ruthann proposed," he said over the phone.

Although Bill had given John permission to discuss the matter with Lacroute, he was instantly depressed by the results, despite the progress. He sat in the corner staring at his feet, muttering and shaking his head. As John continued his review, Randy used the time to work himself into a controlled state of righteous indignation, like a sumo wrestler preparing for a match.

Vinod, fresh from a good night's sleep and a big Indian breakfast, had the clearest head. "The problem is that we have to convince them we're serious about shutting down the company without actually doing it."

Randy had an idea. "We can go ahead and lay off the eighty people tomorrow, and if we don't have an acceptable offer in hand by then, prepay everyone else three weeks' salary, explain what's happening, and send them on their way. The longer EO waits to make a real offer, the more people will have found new jobs."

Robert approved. "Kind of like lighting a fuse on a bomb. They know just when it's going to explode."

To John's dismay, everyone was in favor of this strategy. As the discussion wound around in circles for the next few hours, it became clear that he was the only one willing to use the current proposal as a starting point. Realizing that he had lost control of the situation, he began pacing and talking to himself out loud. "I can't believe this. It is so naive to talk about these kinds of tactics. Those guys are just going to walk out on us and we'll lose everything!" The others ignored him, continuing the conversation. But the frustration got to be too much for him. "That's it. Fine. If you're determined to destroy two hundred jobs tonight, I'm resigning from the board!" John announced. To everyone's surprise, he stormed out of the room — right onto his terrace.

Even under these difficult circumstances, Randy was more than a little amused. "What's he going to do out there?" It was completely dark, and the narrow concrete slab outside the door provided no means of escape.

Robert shrugged. Without a word, John walked back in, crossed the room, and exited through his office door instead. "Should I clock him?" Robert asked.

But there wasn't time. Within ninety seconds, John walked back into the room and flopped into a chair. "All right, I've calmed down. Thanks for your patience."

While the discussion dragged on, the other side cooled their heels, waiting for the group to emerge with something to say. But no one could bring himself to take action. Finally John noticed that Atkinson was heading toward the lobby, and went out to intercept him. He returned a minute later. "There goes the only source of funding for GO and EO! Atkinson is going back to his hotel. He's taking the first flight out tomorrow morning!" Bill walked out quietly to chat with Atkinson.

Though Bill was out of the room, John decided it was time to take action. "I'm calling for a vote," he said. "I move that the board authorize the management team to close a deal on terms equal to or better than the ones on my whiteboard."

Everyone was startled. In the nearly six years since the company was founded, no one had ever formally called for a board vote except for routine matters like approving minutes and stock options. Vinod seconded the motion. But no one spoke up.

Robert was uncertain of the purpose of John's motion, but it seemed innocuous enough. "OK, I guess," he said. Then Vinod and David mumbled their concurrence, mainly so they could get on with the strategy discussion. Now John said "It's approved!"

Just then Bill walked back into the room, and he expressed surprise when he learned of the board's vote. He'd just been chatting with Atkinson; the situation hardly seemed as desperate as John had described. But John's motion sparked Bill into taking some action of his own. "Why don't we propose five million shares and twenty-five million warrants, at fifty cents for seven years. And they have to take over the IBM loans." At least now they had a proposal.

Just as Bernie Lacroute arrived, the others walked into the conference room. Rossmann was clutching a bowl of popcorn to his chest, picking up kernels one by one and stuffing them into his mouth. As Bill outlined the proposal, Rossmann grunted and nodded at each of the figures.

Now it was their turn. Atkinson, Rossmann, and Lacroute disappeared into a separate room for half an hour, then returned to present a final offer. Lacroute wrote it on the board: five million common shares of EO stock and warrants on fifteen million shares, exercisable within four years at one dollar a share. John held his breath as Lacroute moved on to the IBM loan. "We can't accept the IBM liability as it stands. They could call the loan right now and bankrupt the whole company. But we can agree to do two things: take over the loan if you can convince IBM not to call it, and put one million dollars in escrow to cover severance for your people in the event that your negotiations are unsuccessful."

Randy did a quick calculation, then leaned over to Bill. "I'd say that's the equivalent of four to one, around twenty percent," he whispered. His pupils dilated at the thought of turning over responsibility to EO for the shelf full of tangled IBM agreements, which were sitting in his office like a decaying pile of hazardous waste.

Bill looked around the table, focusing in turn on each of the people facing him, searching for one last hint of flexibility in their poker faces. There was Dave Atkinson, who months ago had convinced him to forgo the Reg D offering with assurances that AT&T would help keep the company funded; Bernie Lacroute, who was trying to walk an impossible line between his obligations to EO, KP, and GO; and Alain Rossmann, who had never delivered the

$3 million payment and who, in a fit of pique, had threatened to destroy Penpoint.

Then he turned to his own side of the table. He looked at John Doerr, whose unflagging determination and support over the years had hardened into a determination to save the jobs and project at whatever price they could get; Randy Komisar, whose loyalty to Bill and the team was now the only thing standing between himself and a nice long vacation; and Robert Carr, who watched stoically as the baby he loved so much was about to be handed over to more affluent but less qualified foster parents.

Everyone stared at Bill Campbell in silence as he considered the fate of the two hundred people inside GO and the hundreds of others around the world who had devoted their time and reputations in pursuit of a shared dream. He imagined the human resources team back at the office, working so diligently to ease the pain of those he would lay off tomorrow, not knowing for sure whether they themselves would have jobs at the end of the day. And he thought of me, presumably trapped in an airplane, fighting a losing battle against technology to participate in this crucial meeting.

It was time for him to reveal what only he knew for sure: in the end — when he was confident that he had wrung every last drop out of the other side — he would stop his posturing and gallantly surrender the company, for the ongoing welfare of the people and the project.

"We have a deal," he said softly. Everyone exhaled in relief.

▣

The next day — Friday, August 13, 1993 — was a whirlwind of activity. First thing in the morning, Bill assembled the estaff and the senior managers, to report on the results of the negotiations and to put in place the breakneck schedule he had worked up with Rossmann and his staff. At this meeting, we were to complete the cut list and distribute termination folders to each of the responsible managers. At one o'clock, Bill would send an e-mail message to the company, requiring that all employees assemble in the training room at three. Between those times, the managers would meet one-on-one with the people to be laid off, explain what was happening, and hand them their checks. "Be sure to have plenty of tissues on hand,"

Debbie B cautioned. "Give them a chance to vent if they want." As a veteran of many layoffs at Apple, she understood the importance of a swift announcement and decisive action. "Let everyone know that their voice mail will remain active while they look for a new job, and I'll be setting up a placement center where they can work on their résumés and check job listings."

At three o'clock, Bill would preside over the final comm meeting. There, he and Randy would explain the reasons behind the merger and reinforce that the executive staff was fully in favor of it. Then Bill would announce that both Robert and I would be leaving as well, and give us a chance to speak. When the meeting started, and before the news could leak to the media, EO, GO, and AT&T would all issue simultaneous press releases, focusing on the benefits of the merger and thereby distracting attention from the layoffs. The last scheduled event was at five, when GO's executive staff was to leave the room and Rossmann would address the crowd.

"Remember that people are likely to be pissed," Bill said to the managers. "It's your responsibility to minimize the pain and make this thing come off as smoothly as possible."

Well before three, a line formed outside my door, as word of what was happening quickly spread through the company. To my amazement, people weren't angry at all. Quite the contrary, they were understanding and supportive, prepared to sacrifice their jobs if necessary so the project could continue. For the most part, they just wanted to talk.

"I never suspected we'd be able to last this long without revenues," said Todd Agulnick, our youthful first employee. "But I guess that's just life in the Silicon Valley jungle. By the way, thanks for supporting me while I finished up my B.A. at Stanford. There were a lot more foxes there than at Yale." It took me a moment to realize he was referring to the coeds.

Phil Ydens, the laconic operating systems specialist, came by to let me know he was planning on sticking it out. "I'm not looking forward to working for EO," he said, "but I guess it's just something that has to be done."

During breaks in the flow, I passed the time by calling the investors and bringing them up to date. First I called Scott Sperling of Aneas. "Scott, the way I analyze the numbers, your EO stock is worth seventy-five cents for every dollar you invested in GO, at the very best."

After his support in so many rounds of financing, I expected him to be upset, but he said, "What are you doing about finding new jobs for the people who got laid off?" I told him about Debbie B's outplacement plans.

Next I called Jake Tarr of Arete Ventures. He and his partners had gone way out on a limb for us. But his reaction was the same as Sperling's. "How's the job market out there? If you want to send me some résumés, I'll be happy to circulate them to our other portfolio companies."

Jay Hoag of Chancellor Capital, who by now was a friend of mine, offered comforting words. "Sure I'd rather make money. But bear in mind that this happens all the time. It's a numbers game. Everybody understands that your first responsibility is to take care of your people. Let me know if there's anything I can do."

Last, I put a call in to Norm Vincent of State Farm. "Even though I'll be leaving, I want you to know that the people at EO will continue to support you in every way they can."

Norm was philosophical. "I imagine this is a big disappointment for you, especially after what happened with IBM. But if you ever happen to be in Bloomington, there's a pizza at the Lucca Grill that's got your name on it."

Just before the company meeting, Bill took Robert and me aside. "Guys, this is going to be hard enough, so keep it short and sweet. And whatever you do, don't make everyone cry."

For once, there were no stragglers when Bill called the meeting to order. "I'm sure you all know the news, but I wanted you to hear it directly from the executive staff, and to have a chance to ask questions."

I was watching the crowd, sprawled across chairs, tables, and the floor, paying close attention to Bill's every word. But then I realized that Bill had stopped talking. I turned around to find him so overcome by emotion that he was unable to speak. Powerless to regain his composure, he put his hands to his face and burst into tears. Within seconds, half the people in the room followed suit.

I looked over at Robert, who glanced back at me, suppressing a wry smile. Despite the sadness of the moment, we both saw the ironic humor in it: after all, he was the one who'd admonished *us* about making everyone cry.

Eventually the meeting got under way. After an explanation of the merger by Randy, Robert took over. "I just want to say how sorry I am not to be able to personally see the project through to completion." As he spoke, I heard quiet sobbing from around the room. "It's been a great pleasure working with everyone, and I look forward to seeing you guys set the standard for pen computing with the next version of Penpoint. Whenever I interviewed someone at GO, I told them I wanted them to learn and grow in their job. Part of our culture is to embrace change. And now we have a major change to embrace, and I think in the end it will work out well for everyone."

One of Robert's most endearing qualities was the loyalty and respect he'd earned from his people. Despite his upbeat tone, I could see that for everyone, from his director of applications development to his administrative assistant, losing Robert was like tearing their hearts out.

Then it was my turn. "When I went off to college, I remember wondering how my mother could be so proud and yet so sad at the same time. Now, for the first time, I understand how she felt. Building a company is like raising a child. It goes through phases where it demands your undivided attention, requires your protection, needs your guidance and approval. But eventually it grows up. You have nothing left to teach it, and it goes off to find its own way in the world.

"GO is no longer the tender little startup that Robert, Kevin, and I nurtured so many years ago." I cradled my arms as though rocking a baby. "And we just have to accept that this is a natural part of the corporate life cycle. The goals that all of us have worked so hard to achieve will be better served under the wing of AT&T, as GO reunites once again with our colleagues at EO.

"But I have one last request. Those of us who are lucky enough to be staying with the company through this transition bear a heavy responsibility to those of us who must leave: a responsibility to keep the vision alive, to make sure that all the hard work and sacrifices weren't made in vain. I hope you'll carry forward the ideals and values of our company with pride — treating our partners with courtesy and respect — and go on to find happiness and success in the marketplace."

I looked out at the faces with red, swollen eyes. "And don't forget to write home once in a while!" The room exploded in cheers and applause, which continued until well after each executive staff member had shaken the others' hands and slipped from the room.

The sound echoed as Robert and I walked down the long white hall, then faded like a cloudburst dwindling into gentle rain. Behind us in the distance, I noticed Alain Rossmann standing in the lobby, looking restless as he waited to speak.

"So," Robert said to me, "I guess this is it. I feel like I'm leaving my own funeral."

"I feel like fuckin' Moses," I said, "wandering in the desert for forty years only to be told I can't enter the Promised Land."

"You're starting to talk like Bill," Robert said. Then he smiled. "But I suppose there are worse role models."

He suddenly stopped walking. "Remember that first lunch we had in San Francisco — when you brought that silly leather portfolio and plopped it on the table?"

"Uh-huh."

"You told me that day that we would get one chance to make our point to the world."

"I also said we'd get a shot at the brass ring."

"Yeah, well, maybe we reached a little too far," Robert said. "But I just wanted you to know that I'm not disappointed. I think we got our chance and made our point. We just have to let it go — let other people finish our dream for us."

"Thanks."

After returning calls from the press for several hours, I headed home from my last day at work around eight o'clock, and fell onto the couch. Layne, who was wrapped up in her own professional battles, had just finished eating dinner.

"So, how was your day?" she asked.

Epilogue

THE EO-GO MERGER was formally consummated on January 7, 1994, after a difficult battle to persuade at least 90 percent of the GO stockholders to vote in favor of the transaction. Despite the fact that GO was now gone, the struggle continued.

Of the original GO executive staff, only Mike Homer joined Alain Rossmann's team, feeling that at least one of us should see things through until the next version of the product was shipped. He worked diligently to combine the two organizations and their product plans into one coherent offering. Predictably, the first task was for GO's development teams to complete EO's products and support only AT&T's Hobbit strategy. EO soon stopped development of the Intel version of Penpoint, then halted all work on the Japanese version as well.

Just two weeks after the merger was complete, a senior AT&T executive, Carl Ledbetter, visited EO to discuss a variety of matters with the staff. Ledbetter was forty-four, considered young by AT&T standards. He had been recently hired away from Sun Microsystems to head AT&T's consumer products division, and therefore inherited the responsibility for EO. The meeting turned out to be anything but routine. He informed the EO executive staff that AT&T Microelectronics had decided to kill the Hobbit product line entirely, and so he recommended closing down EO entirely.

As recently as a few weeks before Ledbetter delivered this news, AT&T executives were still pitching rights to the Hobbit to the Taiwanese, as an integral part of the proposed relationship with AT&T and an investment in EO. One of AT&T's slides, entitled

"Why AT&T Microelectronics Is Your Best Long-Term Partner," had as its first bulleted item "AT&T's Corporate Strategy Guarantees Continuing, Large Investment." With the Hobbit gone, it seemed likely that the Taiwan government would no longer be interested in putting money into EO. Since this was a major component of the financing strategy for the company, the prudent course was to shut EO down.

But Rossmann, Homer, Lacroute, Khosla, and Doerr immediately sprang into action. They spent the weekend developing a radical new plan to cut expenses, switch back to Intel processors, and focus future product development on "smart" cellular phones. Unfortunately, this meant the layoff of 100 of the company's remaining 230 employees. Lacroute tracked down Kavner and arranged a Monday evening meeting with him and his staff. Despite the slim chance for approval, the EO–KP team was still hard at work on the proposal on the flight to New Jersey. To everyone's surprise, Kavner and his staff agreed to keep the doors open by providing $18 million in additional bridge financing, pending a final decision at the February 22, 1994, EO board meeting.

But there may have been another reason for AT&T's renewed commitment, not yet disclosed to EO's management: AT&T's potential liability to minority shareholders, including GO's stockholders. According to an internal AT&T memorandum addressed to Kavner, the primary reason AT&T agreed to make the $18 million in interim funding available was to minimize the risk of a legal claim. The memo estimated that the potential damages, if AT&T lost such a suit, could range as high as $100 million.

Rossmann promptly announced to the company that within a week half of the employees would be let go. Their attention was instantly diverted from their work to sending out résumés. Rossmann reluctantly laid off 100 people on February 3.

At the February board meeting, Carl Ledbetter abruptly decided to replace Rossmann with a consultant named Bob Evans. Not long afterward, Rossmann allegedly threatened to take legal action over the matter of his termination, but settled instead for a generous severance.

At the same meeting, the EO board failed for the second time to approve a Memorandum of Understanding to acquire Pensoft, whose product, a personal information manager, was now central to the company's strategy. Having already turned over the software, employees, and assets in good faith, John Zeisler, Pensoft's CEO, was outraged. On May 11 he filed a lawsuit against EO and AT&T for fraud, breach of contract, and conspiracy.

Bob Evans's background was atypical for a Silicon Valley entrepreneur. He was sixty-six years old, and his primary relevant experience was managing large projects at IBM, most notably with the company's mainframe development organization. After thirty-two years with IBM, he joined the investment-banking firm of Hambrecht and Quist. But his distinguished history as an IBM vice president was not the best preparation for the job of managing EO.

The EO executive staff rapidly discovered that Evans thought of engineers more as interchangeable parts in a machine than as specialized cells in an organism. He refused to approve monetary incentives to retain key people, preferring instead to replace them with new hires. As a result, the company lost talented engineers daily. Adding together the two big layoffs — one as part of the merger, the other in February — and the individuals who left on their own initiative, more than 200 years of accumulated experience with the Penpoint code base had walked out the door, leaving only about 150 years at the company to pick up the slack. At GO, the average software engineer had been responsible for 10,000 to 20,000 lines of code, but at EO the load rose to about 80,000. Since the training programs had been dismantled, all newly hired engineers had to learn while on the job, substantially reducing the productivity of the remaining staff.

Evans and his executive staff spent the month of March revising the company's business plan once again before presenting it to the EO board for approval at their meeting on April 7. But just as the staff was finishing up its work, a group of managers from AT&T showed up for a preliminary review. Among them was the strategic advisor to Kavner, whom I had chased from Basking Ridge to a Tokyo sushi bar. He recommended a new strategy for the company — that among other things EO should get out of the hardware business altogether after the current products were completed and concentrate on licensing its software to other manufacturers instead, which was GO's original aim before the merger.

With the company bleeding engineers, the Hobbit project canceled, all of GO's ISVs and hardware licensees gone, and Evans considering reinstating the original GO business plan, Mike Homer realized that the next EO product would never ship. He finally resigned in frustration.

Less than three months later, AT&T decided to withdraw funding for EO completely. On July 29, 1994, AT&T put out a press release, fired nearly all of the remaining staff, and closed the doors for good. A public auction to liquidate the remaining assets was held six weeks later.

GO stockholders never received their EO stock certificates — which hardly mattered, because they were worthless. The only communication of any kind, formal or informal, from AT&T or EO to the original GO stockholders was a terse letter from Bob Evans announcing that the company would cease operations. He added a brief personal postscript: "Those who had the initial vision were correct. It is unfortunate that EO did not succeed to validate that vision."

In looking back over the entire GO–EO experience, it is tempting to blame the failure on management errors, aggressive actions by competitors, and indifference on the part of large corporate partners. While all these played important roles, the project might have withstood them if we had succeeded in building a useful product at a reasonable price that met a clear market need. Just as a coroner may have difficulty assigning a specific cause of death when a person dies of old age, it is hard to point to a single factor that alone was responsible for a company's demise. Like a terminally ill patient, the absence of strong vital signs allowed problems that might otherwise have been correctable to grow unchecked, until the organism fell under their combined weight.

The real question is not why the project died, but rather why it survived as long as it did with no meaningful sales. Without the unflagging efforts of a broad group of supporters, GO might have quietly closed its doors years earlier. The project's longevity is a testament to the force of will and the compelling power of belief in an idea. It is comforting to know that the human impulse to make the world a better place is not today confined to the young or the foolish. It is alive and well among the people that live and work in Silicon Valley.

After leaving Apple in October 1993, John Sculley took over as CEO of Spectrum Information Technologies, a small East Coast company. Shortly thereafter he resigned, after some preexisting irregularities were uncovered.

In June 1994, Bob Kavner resigned from AT&T to take a position at Creative Artists Agency, the largest Hollywood talent agency, where he was to head its efforts in multimedia entertainment.

After taking up John Doerr's offer of temporary office space at Kleiner Perkins, Bill Campbell was asked to become CEO of Intuit, the maker of the popular Quicken financial software. Intuit soon agreed to be acquired by Microsoft.

Randy Komisar became CEO of LucasArts Entertainment Company, a manufacturer of computer games, owned by George Lucas, the director best known for his *Star Wars* films.

Robert Carr accepted a job as vice president of Core Technology, developing new products for Autodesk, the leading provider of computer-aided design software.

After IBM acquired his company, Metaphor, David Liddle founded a Silicon Valley laboratory called Interval Research with the backing of Paul Allen, the other founder of Microsoft. Kevin Doren later joined the laboratory's staff of respected scientists.

⊡

After the FTC deadlocked, 2-2, on whether to take action against Microsoft and officially ended its investigation, the Justice Department announced, in August 1993, that it would initiate a similar antitrust probe of Microsoft. In July 1994 the Justice Department and Microsoft agreed to a consent decree, and the department filed an antitrust complaint alleging that Microsoft used exclusionary and anticompetitive licensing agreements to market certain of its software products and thus had unlawfully maintained its monopoly of personal computer operating systems. The consent decree specifically prohibited Microsoft from requiring its licensees to pay a "per processor" royalty on every product using an Intel chip, whether or not it uses a Microsoft operating system. The decree also prohibited Microsoft from entering into contracts that restrict PC manufacturers from licensing, selling, or distributing products using competing operating systems.

As of December 1994, Apple's Newton product line had produced only lackluster sales and was widely regarded as having fallen short of expectations. General Magic had just launched its first version of Magic Cap, in conjunction with AT&T. Microsoft announced that it had canceled Winpad, its unreleased successor to Pen Windows, after Intel dropped its plans to produce Polar, its response to AT&T's Hobbit chip. The Winpad team was folded into the Windows group, to start over yet again.

State Farm had completed a field test of its first Penpoint-based claims-estimating application. Its preliminary studies indicated that this one application could save State Farm as much as $85 million a year, and so Norm Vincent was preparing to order $50 million worth of pen computers. Seeing no future for Penpoint, he was finally persuaded to switch to Pen OS/2.

After a few days of relaxing at home, I began to think about what to do next. Alan Fisher, an old friend of mine from my time at Teknowledge, called to see if I had any thoughts about how he could leverage some interesting technology he had recently developed. His team of software engineers had just completed a contract with Charles Schwab, the discount brokerage house, to build a Windows front end for its portfolio-tracking and order-management system, called Street Smart. "Basically, customers can dial up the Schwab data center from their home computers, update their portfolios, get stock quotes, place buy and sell orders, and browse through the latest news about companies and products," he explained.

I called him back the next day. "You know, it seems to me that buying and selling stocks is a lot like buying and selling anything else — it's just much more efficient. When you think about it, why should there be a fixed price for anything, whether it's stereos or cereal? I wonder what would happen if we put together the ultimate electronic market for consumer goods."

He was intrigued, but had his doubts. "Electronic retailing has had limited success at best. It's horrible, just boring lists of items and prices."

"But suppose you did it differently," I said. "You could build an interface that was engaging and attractive, with lots of pictures and product information. You make it free to the customers and just take a percentage of each sale. But most important, you let the prices float in response to supply and demand just like the stock market."

"You know what else," he said. "Since we could see which items a person was looking at, and for how long, we would get information on their interests before they bought anything. Then we could shoot them an electronic coupon — say five bucks off on a pair of skis they were examining — but it's only good for the next thirty minutes."

We both paused to imagine the possibilities. Alan was the first to speak. "Do you realize what this means? This could be a whole new form of retail, a whole new distribution channel for goods, with optimal pricing and infinite shelf space. A billion-dollar industry."

I felt that wonderful sense of epiphany again — but I didn't know if I had the energy to blindly follow my heart this time. "Christ, Alan, I'd like to take a break before jumping into something like this right after GO."

"Yeah, I heard it was a pretty wild ride."

"I'm telling you, no one would believe it."

"Maybe you should write a book about it. *Then* we can get started."

Which I did, and it ends with this sentence.

AUTHOR'S NOTE

CHRONOLOGY

APPENDIX

GLOSSARY

INDEX

AUTHOR'S NOTE

From the founding of GO in 1987 through its acquisition by AT&T in 1994, I kept a thorough diary with the idea of one day writing a book about the experience of starting a new company. For the first two years — until the time we moved the office from San Francisco to Foster City — I was able to find a few hours each weekend to sit down and write up the week's important events. After the move, when GO started growing quickly, this proved impossible, so I bought a pocket tape recorder and kept an oral commentary instead, making recordings once or twice a week while commuting to work. Periodically, I sent off a completed tape to a typing service. The transcribed notes began arriving in large manila envelopes, which I stuffed into a drawer at home.

When I transferred the contents into binders, I discovered that the approximately twenty hours of tape comprised almost a thousand pages of detailed notes. This running chronicle was supplemented by more than fifty megabytes of incoming and outgoing e-mail, memos, budgets, reports, minutes, proposals, and performance reviews. In addition to this electronic record, I collected appointment calendars, correspondence, brochures, newspaper clippings, drawings, prototypes, pictures, and videotapes. It was no secret to GO's board of directors and employees that I was gathering information for a book project, and many of them helped by donating important documents.

The careful reader will notice that I was not present for several scenes in the latter part of the book. To reconstruct these episodes, I relied on the taped recollections of as many of the participants as possible. I am deeply indebted to several people — especially Robert Carr, Bill Campbell, Randy Komisar, and John Doerr — who gave freely of their time to describe these scenes.

From these sources and my own recollections I selected the small fraction of details and incidents that best represented the courage and imagination, frustration and humor inherent in a business venture. I made every effort to ensure accuracy, though in many places I had to reconstruct dialogue that I felt was true to the meaning, intent, and style of the speakers.

I am grateful to John Sterling, my editor, who kept the story and pace on track, and particularly to Larry Cooper, my manuscript editor, who tuned up the prose and kept it running smoothly. My tireless agent, Kris Dahl, and her dedicated assistant, Dorothea Herrey, explained the inner workings of the publishing industry and provided invaluable feedback on interim manuscript drafts. Thanks also to Betsy Peterson for the legal review and to Joseph DeRuvo, Jr., for permission to reprint his letter to me. I am indebted to my wife, Layne, who provided her usual sage advice, and to our new baby girl, Lily, who was kind enough to wait until the first draft was complete before joining us.

From this experience I learned three lessons: real life is infinitely more subtle and complex than what you can fit in a book; writing is more fun than reading, and easy work if you can get it; and no matter how much you warn people in advance, some of them still get upset when they see in print what they actually said and did. I apologize to all those who worked so hard for so many years and weren't mentioned in the book. I also apologize to all those who worked so hard for so many years and *were* mentioned in the book.

If you'd like to contact Jerry Kaplan regarding *Startup,* please send your e-mail messages to kaplanj@hmco.com.

CHRONOLOGY

···

FEBRUARY 1987 Mitchell Kapor and Jerry Kaplan discuss the idea of a pen computer during a cross-country flight.

AUGUST 1987 GO is formally incorporated.

JULY 1988 Bill Gates of Microsoft is shown a deskbound prototype of GO's pen computer.

MARCH 1989 GO presents its plans to the Research Board, a group of the top information systems executives in the country.

JUNE 1989 GO demonstrates its pen computer to State Farm, attempting to win the insurer's business over IBM, Hewlett Packard, and Wang.

MARCH 1990 After learning of State Farm's decision to work with GO, IBM agrees to license GO's Penpoint operating system.

JULY 1990 GO announces a partnership with IBM.

JANUARY 1991 GO's Penpoint "developer's release" is announced to broad computer industry acclaim.

MARCH 1991 GO and Microsoft face off publicly for the first time at the PC Forum in Tucson.

APRIL 1991   Jerry Kaplan briefs Federal Trade Commission staffers who are investigating possible antitrust violations by Microsoft.

JUNE 1991   GO and Microsoft both present their operating systems at the NCR3125 pen computer announcement in New York City.

JULY 1991   GO agrees to sell its hardware design group to AT&T and other investors. EO, an offshoot of GO, is formed to build pen computers based on Penpoint.

OCTOBER 1991   IBM begins promoting Pen OS/2 as an alternative to Penpoint.

APRIL 1992   Penpoint version 1.0 is released. IBM announces its Thinkpad pen computer.

MAY 1992   John Sculley discusses Newton, Apple's "personal digital assistant," at the Consumer Electronics Show in Chicago.

JULY 1992   GO announces a partnership with AT&T. Apple demonstrates Newton at the Mobile92 conference.

NOVEMBER 1992   EO and AT&T demonstrate the EO440 personal communicator at COMDEX.

FEBRUARY 1993   General Magic announces a partnership with AT&T, Sony, Motorola, Matsushita, and Phillips to promote Magic Cap.

JUNE 1993   AT&T buys a majority position in EO.

AUGUST 1993   After AT&T considers dropping GO's Penpoint for Apple's Newton, GO agrees in principle to sell the company to AT&T by merging with EO.

JANUARY 1994   The EO–GO merger is completed.

JULY 1994   AT&T decides to close down EO.

APPENDIX

....

The following is the full text of an e-mail message sent by Bill Gates to his senior staff soon after the meeting at GO described on pages 64–67. (E-mail addresses have been deleted by the author.)

From: billg Thu Jul 14 16:42:49 1988
To: [6 names deleted]
Subject: GO corporation
Cc: [3 names deleted]
Date: Thu Jul 14 16:42:45 1988

Jeff Harbers and I met with Jerry Kaplan and Bob Carr of GO corporation Monday afternoon.

Basically they are building a machine that Kay and I talked about building a long time ago — a machine with no keyboard and no disk using static memory. It's like an 80286 version of the model 200 with 2meg–8meg using a writing stylus with handwriting recognition for input. According to Marquardt there are a few other people building things like this — in fact there was one discussed in the WSJ this week. It's notebook size. The LCD is 640x400 so about 55DPI (which I don't think is enough).

They are doing ALL their own system software — a protect mode OS for 286 using visual objects (like everyone!). It's multitasking. The interface metaphor is a set of named folders with tabs on the right hand side each containing any number of numbered pages and each page has on it just ink (writing) or rectangles that contain application sessions (which can be zoomed/unzoomed). All the old ideas like using gestures for various commands they have "redis-covered." They will announce in 1990 at $3k. Modem is optional. They will bundle some drawing/wp/filing/notetaking/mail software but they want to get third party ISVs including Microsoft.

We tried out their handwriting stuff and it was terrible. It's very possible to do this stuff correctly and maybe they will but they haven't yet.

ANALYSIS: This machine should be built as an open standard by a bunch of Japanese makers. The software layers should be more compatible with desktop stuff. Kaplan isn't the best CEO. They have some OK ideas but I don't think this thing will be big. We do need to think about note taking and the fact that small machines can be used everywhere especially with this input approach but I don't think we should be an ISV for them.

ACTION ITEMS: Gregw — Carr wants our debug format to do a remote debugger. If it is written up and easy to send then send it to him with a letter saying they will use it for developing on their machine only. If it is hard then have someone call and say sorry.

Joachimk/Steveb — we should be selling system software to people like this. He is fairly far along at this point. What would we sell him? Either stripped down PM or WINDOWS. WINDOWS is the best choice I think. Tell him to use extended memory. He won't like this but it will sure help him with applications developers. Who can take a pass at this with Carr?

Mikemap — Another applications opportunity but unless we want something that fits on this machine for the desktop I doubt it makes sense. They do want to create connections between stuff on their machine and popular desktop stuff so we want to be friends with them even if they are not an ISV. They offered to come up and present their concepts to a larger group up here. I doubt that makes sense. I will talk about this class of machine in our Saturday morning meeting.

GLOSSARY

API (application programming interface). The functions that one program makes available for other programs to use. Or, more specifically, the particular way in which these functions are made available. In the case of an operating system, like Windows or Penpoint, the API comprises numerous functions that the programmer can access to perform a variety of operations within that system, such as reading or writing data, managing memory, and displaying information inside a window on the screen.

ASCII. A standard binary code for letters, numbers, and special symbols developed by the American Standards Committee which can be recognized by all computer programs.

ASIC (application-specific integrated circuit). A computer chip designed for a specific application. ASICs are often developed by a company for use in one or more of its own proprietary products, in conjunction with standard off-the-shelf parts.

Assembly language. A low-level programming language whose structure closely corresponds to the logical design of the computer, often translating easily into simple binary instruction codes.

BIOS (basic input/output system). The low-level program, particularly on IBM-compatible PCs, that manages data going into or out of the central processing unit (see *CPU*).

Buildmaster. A software engineer who assembles the various components of a large software system into a version of the completed system.

CMOS (complementary metal-oxide semiconductor). A specific type of computer chip that consumes less power than other types of chips and so is often used in devices that run on batteries.

Common stock. In a startup, the stock usually held by employees and others who provide a service to the company.

CPU (central processing unit). The part of a computer's circuitry that carries out the instructions encoded in software. Program instructions and data are loaded from memory into the CPU, which then performs the indicated operations on the data.

DOS (disk operating system). Until recently, the most widely used operating system for IBM-compatible personal computers. DOS was developed by Microsoft Corporation and licensed to IBM (see *Operating system*).

EGA (extended graphics array) **controller**. A particular computer chip that controls the display of information on the screen of a personal computer.

EPROM (erasable programmable read-only memory). A type of memory that retains its contents even when the computer's power is turned off, but that can be erased with ultraviolet light and reprogrammed. EPROMs are often used in hardware prototypes that require frequent changes.

Flip-flop. The common name for a simple digital circuit element that can be switched on and off.

Gestures. Symbols drawn with an electronic pen that are interpreted as commands by a pen computer. For example, in Penpoint software the gesture *X* over a word represents a command to delete the word.

GOLD chip (GO logic device). An ASIC chip designed by GO hardware engineers to perform specific functions required by GO's prototype pen computers.

IPO (initial public offering). A financing event whereby a private company, often a startup, "goes public" by selling shares of stock to the public and is subsequently listed on a stock exchange.

ISV (independent software vendor). A software developer that writes or sells programs based on another company's programming system. The program is usually an application that will run on an operating system.

Limited partners. Investors in a project who provide money but do not have responsibility for managing the enterprise, and so have limited liability for the actions taken on behalf of the project.

Market capitalization. The total value of a company, calculated by multiplying the number of outstanding shares by the value of those shares.

Megabyte. Approximately one million (1,048,576) bytes. A convenient measure of memory size; one megabyte equals 2^{20} bytes.

NDA (non-disclosure agreement). A legal agreement in which a company (or individual) that discloses confidential information to another is assured that the other company (or individual) will limit its use of the information to certain stated purposes.

Operating system. The fundamental program that operates a computer. It provides access to input/output devices, manages system resources such as memory and files, and controls the execution of applications programs. Modern operating systems, such as Windows and Penpoint, provide a consistent "look and feel" to the user by offering functional and stylistic guidelines for applications designers.

Personal communicators. Portable computers used for one or more forms of communication, such as telephoning, faxing, and paging.

Pixels (picture elements). The small dots out of which digital pictures are composed; also a unit of measurement used to identify a specific position on a computer screen.

Post-money valuation. The market capitalization of a company immediately after a financing event. The post-money valuation is equal to the pre-money valuation plus any newly invested cash before it is spent.

Preferred stock. One or more special classes of stock issued to investors in a startup, as opposed to common stock, which is issued to employees. Preferred shareholders get certain negotiated privileges (called preferences), such as the right to get their original invested money back before the common shareholders do if the company is sold or liquidated for less than the price they paid for their stock.

Pre-money valuation. The market capitalization of a company immediately before a financing event — i.e., the value that new investors are willing to place on the company prior to making their investment.

Propagation delay. The length of time it takes for a voltage to become stable and observable at the termination point of a circuit — i.e., the time it takes before a digital signal can reliably be read.

Pro-rata share. The part of a financing that an existing shareholder would buy if all previous investors take part in the financing in proportion to their current holdings in the company. If all investors take their pro-rata shares, then their relative percentage ownership in the company will not change as a result of the financing.

Regulation D. A securities regulation that allows a private company to sell stock to a limited number of individual investors without the restrictions and reporting requirements governing public corporations.

Restricted stock. Stock that cannot be sold except under certain circumstances, such as at an initial public offering (see *IPO*) or after first offering to sell the stock back to the company.

RFP. A Request for Proposal, which usually contains written specifications of a complex system that suppliers can then bid on.

RISC (reduced instruction set computer). A kind of computer design that employs fewer and simpler basic instructions, which can be executed more quickly and efficiently.

ROM (read-only memory). A memory chip whose contents are stored permanently when it is manufactured, and from then on can only be read, not changed. ROMs are usually less expensive than other, more flexible memory chips, and they retain their contents when power is turned off. ROMs can be used to store fixed programs, such as an operating system, which must always be present in the memory of a computer.

Software build. The result of the process by which a software engineer assembles the various components of a large software system into a version of the completed system.

Software interrupts. A signal generated by a computer's software to suspend temporarily the normal flow of instructions through the CPU and perform some urgent task, such as fetching data from a disk, reading the location of an electronic pen, or moving an image on the screen.

Source code. The "symbolic" form of a program, which allows a programmer to make substantive changes in the functions of that program.

SRAM (static random-access memory). A type of memory chip that can retain its contents with only a small amount of power. SRAM memory is useful in portable devices, because the memory can be kept alive for long periods without draining the battery.

Vesting period. The time during which employees or other service providers must perform work for a company in order to gain clear title to their stock. In startups, the vesting period is typically four years.

Warrant. A right to buy stock at a specified price for a given period of time, called the term. The holder of the warrant will usually exercise this right if the stock value rises above the specified price during the term. If the stock value stays below the specified price, the holder will likely let the warrant expire.

INDEX

ABC News, 178, 214

Active Book, 202–4, 261

ADP, 219

Advanced Micro Devices, 279, 280

Agenda (conference), 158

Agenda. *See* Lotus Agenda

Agulnick, Todd, 38, 49, 51–52, 57–58, 74, 93, 94, 287

Akers, John, 88

Allen, Bob, 239, 264

Allen, Paul, 295

Alsop, Stewart, 84–85, 158, 168, 176, 227

Aneas (Harvard Management), 69, 78, 220, 287

Apple: alliance with IBM, 201, 210–12; competition with, 19–20, 47–48, 156–57, 227–28, 246, 267, 274, 277–78, 295–96; hardware engineering at, 16; in lawsuit with Microsoft, 181

Apple Newton personal digital assistant (PDA): ARM chip in, 156, 202, 234; AT&T's interest in, 264; IBM's interest in, 210–11, 274; John Sculley's promotion of, 156–57, 227–28; lackluster sales of, 295–96; as Penpoint's major competitor, 47; prototypes of, 228, 232, 234, 246; sales training for, 277–78

Apple Powerbook, 265

Application programming interfaces (APIs), 49–50, 144–45

Arete Ventures, 71, 288

ARM (Acorn RISC Machine) chips, 156, 202–3, 234

Artificial intelligence, 2

ASCII (plain text), 25

ASCII Corp. (Japan), 67

Ashton-Tate, 29, 30, 63, 64, 67

Associated Press (AP), 150

Atkinson, Dave: and AT&T's investment in GO Corp., 248, 255, 266–67; and EO–GO merger, 266–67, 271, 274–76, 280, 282–85; and formation of EO Corp., 207–8; and reorganization of EO, GO, and AT&T, 260–63

AT&T: corporate headquarters of, 258–59; financing of EO Corp. by, 241, 261–62, 292; financing of GO Corp. by, 248–49, 251–52, 255; partnership with GO Corp., 202–3, 217–18, 231, 232–34; promotion of Penpoint and EO

AT&T (*cont.*)
    440 by, 244–46; as proposed
    buyer of GO Corp., 255, 257–60;
    reorganization of partnership
    with GO and EO, 260–67
AT&T Bell Labs, 156
AT&T Microelectronics, 233, 260,
    291–92
AT&T Personal Communications
    Systems, 233
AT&T Telepen RFP, 217, 230, 232,
    239
Auction, to liquidate EO Corp., 1,
    294
Autodesk, 295

Ballmer, Steve, 158–59, 214–15
Bank of America, 91
Baranski, Celeste, 39–40, 48, 49,
    51, 93, 102–3, 107–11, 161, 208–9
BASIC, 11
Baum, Michael, 165, 177, 194–95,
    250
Bechtolsheim, Andy, 69
Be Labs, 47
Belove, Ed, 7
Bessemer Partners, 72, 76, 81
Biondolillo, Debbie, 273, 281, 286,
    287
BIOS, 187–91
Boston Computer Society, 161, 168
Božinović, Rasha, 74–76, 80–81
Broadbent, Carol, 149–50, 154–55,
    160–62, 164–67, 185, 187
Byers, Brook, 25, 26, 243
Byte, 168

Campbell, Bill: as CEO of GO
    Corp., 159, 168, 196–97, 198, 200;
    at Claris, 43, 159; and competi-
    tion for Penpoint, 186, 197, 199–

200, 201, 204, 253–54; and EO–
    GO merger, 270–71, 274–77, 278–
    81, 282–86, 286–87, 288; and GO
    Corp.'s financial problems, 216–
    18, 248, 273–74; and GO Corp.'s
    partnership with AT&T, 233,
    238–40; and GO Corp.'s partner-
    ship with IBM, 210–12; at Intuit,
    295; and problems with EO Corp.,
    241, 250–51; and reorganization of
    GO, EO, and AT&T, 262–67; and
    strategy review, 229–32
Cannavino, Jim, 245; and IBM's
    deal with Apple, 210–11; and ne-
    gotiations over IBM partner-
    ship, 124, 127, 128–30, 134–35,
    136, 141–42, 162; and press an-
    nouncement, 145, 148–49; selling
    of Metaphor to IBM, 201
Carr, Robert: and competition
    from Newton, 227, 228; demon-
    strations of Penpoint by, 174–75,
    226–27, 245; demonstrations of
    prototypes by, 61–64, 110, 112–14;
    and EO–GO merger, 281, 282–
    86, 288–89, 290; at first working
    session, 33–36; and management
    of GO Corp., 38, 49–50; and
    problems at EO Corp., 240; and
    proposed sale of GO Corp., 254–
    55, 257; recruitment of, 29–30;
    and software development, 102,
    105–6, 200, 224, 253
Chancellor Capital Management,
    69, 73, 288
Charles Schwab (brokerage), 296
Claris, 43, 159
Clean room BIOS, 187–88
CNN (Cable News Network), 235
COMDEX, 12–13, 82–85, 127–28, 213–
    16, 218, 244–46

Communications Intelligence
   Corp., 230
Compaq, 8, 9, 253–54
Consumer Electronics Show, 228
Corporate culture, 37, 117
Corporate law and lawyers, 18, 37,
   120, 177–79, 220–21
Cravath, Swain & Moore, 121
Critter P. Spats, 7, 31–33, 115
Croll, John, 178–79, 180, 182–85

Department of Justice, U.S., 120,
   183, 295
DeRuvo, Joseph, 205–6
Doerr, John: advice about IBM
   partnership deal, 130, 143, 163; at
   Agenda, 159–60; appointment to
   GO Corp.'s board, 95; at COM-
   DEX, 83, 84–85; description of,
   28; and EO–GO merger, 275–77,
   282–85; as financial backer for
   GO Corp., 23, 24, 26, 27–29, 61–
   64, 68–69, 72, 73, 76, 78–80; and
   GOLD chip, 107; and management of GO Corp., 157; and ongoing financing of GO Corp.,
   150–51, 152, 247–48; and problems at EO Corp., 242; and proposed sale of GO Corp., 264,
   265; reorganization of EO Corp.
   following merger, 292; and
   RISC architecture for Penpoint,
   201, 202–4
Doren, Kevin: and demonstrations of prototype, 61–64, 71; departure of, 240–42; at first working session, 33–36; and GO
   Corp.'s partnership with AT&T,
   234; and hardware development,
   103, 107–11, 208; at Interval Research, 295; and licensing of key

technologies, 223; and product
   launch, 166, 167; recruitment of,
   29, 30–33
Doren, Melanie, 31–33
DOS, 200, 212, 236
Dvorak, John, 83, 84, 85
Dyson, Esther, 173, 175

Earthquakes, 125–27
Electronic ink, 13, 42, 56, 57, 177
Electronic pens. See Pen computers
Electronic tablets, 14
Entrepreneurship, 21–23
EO Corp.: auction of materials
   from, 1, 294; business plan for,
   293; closing of, 291–94; competition for, 234; financing of, 241,
   261–62, 292; layoffs at, 292, 293,
   294; management problems at,
   240–42, 250–51; merger with GO
   Corp., 266, 269–77, 280–81, 282–
   86, 291; reorganization of, 260–
   67; software development at,
   241, 250, 260, 293; startup of,
   206–9; valuation of, 261, 271–76
EO 440, 244–45
Esber, Ed, 64, 67
Evans, Bob, 292–94
Exley, Charles, 185

Feature creep, 253
Federal Trade Commission (FTC),
   176, 177, 178–79, 181, 182–85, 295
FFS2 (Flash File System 2), 223–24
Financing: from AT&T, 248–49,
   251–52, 255; changes in, 54; following IBM deal, 148, 150–53,
   157, 211–12; from IBM, 118–19,
   142–45, 211–12, 217, 225–26, 271,
   285; initial, 1, 18, 20, 22–29, 59–
   81; Kapor's assistance with,

Financing (*cont.*)
18, 20, 23, 24, 28, 61–64, 72, 94;
and revenue, 93, 254, 287, 294;
by sweat equity, 22; Taiwanese,
255–56, 263, 267; total amount
of, 1; by venture capitalists, 11–
12, 20, 22–36, 59–81, 213
Fisher, Alan, 296–97
Fisher, Ron, 189, 190
Flat-panel displays, 56
Foundation for Educational Software auction, 158
Frink, Lloyd, 105, 175, 177, 182, 223,
235, 236–37, 253
Furneaux, Jim, 72–73, 76–78

Gassée, Jean-Louis, 19, 43–47
Gates, Bill: and competitive strategies at Microsoft, 16, 63, 64, 67,
101, 135, 176, 180–82, 199, 305–6;
promotion of Windows projects
by, 44, 214–15
General Magic, 218, 230, 252, 296
"Glue" ASIC, 107
GO Corp.: auction of materials
from, 1, 294; board of directors
of, 28, 58, 60, 94–95; company
mascots, 50–52; company-mission statement for, 72; corporate
rhythm of, 196–97; executive
staff of, 28, 29–36, 51, 56–57, 78–
80, 217; financing of, 1, 18, 20,
22–29, 59–81, 118–19, 142–45, 148,
150–53, 157, 211–12, 216–20, 217,
225–26, 243–44, 248–49, 251–52,
255–56, 263, 267, 271, 285; first
working session of, 33; formal
business plan for, 68, 294;
founding of, 2, 18–20; goals of,
18, 26–27; incorporation of, 36;
initial development team for, 17–

18; initial public offering (IPO)
for, 219, 248; insurance for, 72;
layoffs at, 216–17, 271, 281, 286,
287; licensing agreements from,
165–66; logo for, 154–55; management principles at, 48–49, 101;
marketing by, 149, 247; merger
with EO Corp., 266, 269–77,
280–81, 282–86, 291; naming of,
18, 29; offices for, 17, 18, 38, 41,
92–93, 201, 253; partnership with
AT&T, 231, 232–34; partnership
with IBM, 115–16, 117–45; parts
suppliers for, 102–3, 107; potential customers for, 83, 87; preparations for merger with EO
Corp., 286–90; product development schedule at, 148, 160, 252–
53, 256; proposed sale of, 254–
57; publicity for, 145, 146–69,
235; reorganization of, 260–67;
revenue of, 93, 254, 287, 294;
server for, 126; staff for, 38–41,
74–76, 201, 217, 271; State Farm
Request for Proposal (RFP) to,
100–101, 105, 106–16; strategy review for, 229–32; technical team
for, 39–40, 92; valuations of, 70,
71–72, 73–74, 79–80, 93, 151, 219–
20, 247–48, 271–76
GOLD chip, 107–8, 110
Grid Corp., 40, 165, 180, 246
Grove, Andy, 69, 151–53, 236

Hammel, Paul, 208–9
Handwriting-recognition software:
and ASCII text conversion, 16,
25, 26; consultation with experts, 18, 72, 74–76; IBM research on, 149; importance of,
16; Microsoft research on, 181–

82; problems with, 56, 57–58; speed of, 72

Hansche, Lance, 189–90

Hardware development, 35, 56, 93, 102–3, 107–11, 161, 208–9

Hauser, Hermann, 203, 204, 207–8, 261

Hearst, Will, 69, 84, 85

Hewlett Packard, 98, 100–101, 116, 118, 119

Hillis, Danny, 88

Hoag, Jay, 73, 288

Hobbit RISC chips (AT&T), 156, 202–3, 206–9, 233, 240, 250, 260, 283, 291

Homer, Mike, 230; comparison of Penpoint to Newton, 227, 228; at EO Corp., 291, 292, 294; and GO Corp.'s partnership with AT&T, 232, 248; and Newton sales training, 277–78; and lay-offs at GO Corp, 217; recruit-ment of, 198–99; and switch to RISC chips, 199–200

IBM: acquisition of Metaphor by, 295; alliance with Apple, 201, 210–12; competition from, 225; contract with, 120–24; corporate culture of, 117; corporate head-quarters for, 133–34; delivery schedule to, 122; financial prob-lems at, 143, 231; loan from, 211–12, 217, 225–26, 230–31, 271, 285; mainframe computers, 98, 118, 120; management committee of, 118; partnership with GO Corp., 115–16, 117–45, 180, 210–13; pen computers and software from, 212, 230, 252; proposed acquisi-tion of GO Corp. by, 129–31,

134; State Farm Request for Pro-posal (RFP) to, 100–101, 106, 114–15; T J. Watson Research Labs, 118

IBM PS-LOB (personal systems line of business), 124

IBM Thinkpad, 225, 226, 231, 252

Independent software vendors (ISVs), 63, 165, 166, 200, 213, 215, 216, 227, 230, 260

InfoWorld, 226

Initial public offering (IPO), 13, 23, 65, 219, 248

Intel Corp., 8, 39, 44, 151–53, 199–200, 223, 235–38. *See also* Micro-processor chips

Intellectual property: as collateral for deal with IBM, 121, 123–24, 142, 144–45; protection of, 22, 63, 65, 102

Interval Research, 295

Intuit (company), 295

Investment banks, 69, 219, 243

Japan: GO Corp. office in, 253; parts suppliers in, 103–5

Jobs, Steve, 20, 44–45, 69, 88, 154, 155

Johnson, Al, 118–19, 121

Joshi, Dr., 2–3

Joy, Bill, 69

Kalb, John, 121, 122–23, 127, 133–36, 142, 162–63

Kaleida's Script-X, 211

Kaplan, Jerry: activities following termination at GO Corp., 296–97; as CEO of GO Corp., 28; and demonstration of proto-type, 61–64; description of, 8; di-ary maintained by, 298; at first working session, 33–36; last days

Kaplan, Jerry (*cont.*)
  at GO Corp., 286–90; marriage
  of, 158, 244; parents of, 52–56,
  131–33, 136–41; personal goals of,
  26–27; Ph.D. work of, 2–3, 6, 24;
  and product launch, 166–67; on
  publicity tour, 149; reaching
  burnout, 157; recruitment of
  Carr and Doren by, 29–33;
  search for willing buyer for GO
  Corp., 256, 257–60
Kaplan, Layne, 158, 162, 172–73,
  228, 244, 290
Kaplan, Mickey, 52–54, 132, 136–39,
  140
Kaplan, Murray, 52–56, 131–33, 136–
  41, 268
Kapor, Mitchell: and ON, 18, 29;
  changing values of, 272–73; de-
  fection of, 20; description of, 7;
  as financial backer of GO
  Corp., 28, 61–64, 72, 94; and
  founding of GO Corp., 17–18;
  personal jet of, 4–5, 7, 272; as a
  prolific note taker, 9; research in
  artificial intelligence by, 5–6;
  start of Lotus by, 11–13
Kavner, Bob, 203–4, 207, 238–39,
  248, 258, 261, 262–67, 269, 270,
  292, 295
Kaye, Gary, 178, 214
Keyboards, elimination of, 10–11,
  13, 30
Khosla, Vinod: at Agenda, 159–60;
  and EO–GO merger, 282, 283,
  284; as financial backer of GO
  Corp., 26, 28, 61–64, 72; and negotiation of IBM partnership
  deal, 142–43, 163; and GO's ongoing financial problems, 248;
  and proposed sale of GO Corp.,

265; reorganization of EO Corp.
  following merger, 292; and
  RISC architecture for Penpoint,
  201, 202–3
King, Sue, 119, 121, 122, 127, 141,
  142, 148, 153, 157, 231
Kinser, Dave, 274, 276
Kleiner Perkins Caufield & Byers
  (KP), 23–27, 28, 78, 151, 200, 203,
  261, 274–76, 281, 282–86; limited
  partners of, 68–69
Knowledge Navigator, 46
Komisar, Randy, 230; and EO–GO
  merger, 274–76, 279–81, 282–86,
  287, 288; and financing from
  AT&T, 206–9; and GO's financial problems, 217, 247–48, 255,
  256, 273–74; hiring of, 198; and
  IBM loan, 210–12, 225–26, 230–
  31; at LucasArts, 295; and problems with EO Corp., 250; and
  proposed sale of GO Corp., 255;
  and switch to RISC chips, 200–
  204

Lacroute, Bernie, 203–4, 206–9,
  240, 256, 261, 263–67, 277, 282,
  283, 285, 292
Lamond, Pierre, 70
Laptop computers. *See* Portable
  computers
Lead investors, searches for, 60,
  73, 74, 76, 79–80, 81, 150–51, 220
Ledbetter, Carl, 291, 292
Legal counsel. *See* Corporate law
  and lawyers
Liddle, David: attempt to save
  Metaphor, 201; and EO–GO
  merger, 282, 284; on GO Corp.
  board, 95, 124–25, 127, 143, 162–
  63, 240, 241; at Interval Re-

search, 295; marriage of, 272; and GO's ongoing financial problems, 248–49; and proposed sale of GO Corp., 265
LISP computers, 6
Lotus Agenda, 6–7, 9, 17, 27, 165
Lotus Development Corp., 4, 5, 11–13, 63, 64, 67, 255
Lotus 1-2-3, 12

Magic Cap (General Magic), 252, 296
Mainframe computers, 14, 98, 118, 120
Manzi, Jim, 64, 67
Market research, 70, 72, 105
Markoff, John, 242
Marquardt, Dave, 71
Mason, Marcia, 186–87, 192, 194
Matsushita (Japan), 202, 203, 264
Maxwell, Holli, 204–6
McNealy, Scott, 69
Merrill Lynch, 91
Metaphor (company), 95, 124, 201, 296
Metro Center (Foster City), 92
Miami Brain Bank, 138
Microprocessor chips; by Intel, 8, 39, 152, 253; RISC chips, 156, 200, 202–3, 206–9, 233, 234, 240, 250, 260, 283, 291
Microsoft Corp.: acquisition of Intuit by, 295; and applications software for pen computers, 63, 64–67, 101–2, 105–6; competition with, 16, 44, 160, 170–95, 198, 213–16, 237–38; cross-licensing with IBM, 122–23, 135; and FFS2 (Flash File System 2), 223; FTC investigation of, 176, 177, 178–79, 181, 182–85, 295; initial public of-

fering (IPO) of, 65; in lawsuit with Apple, 181; pricing arrangements of, 179–80; and RISC chips, 253; and telephone interface standards market, 218; Winpad, 253, 296
Miller, Peter, 17, 18, 20, 29, 156
Minicomputers, development of, 14
Mobile92 conference, 232
Mohan, Alok, 186
Monitors, EGA controller chips for, 103, 108
Moon, Rev. Sun Myung, 104–5
Motorola 68000, 39

NCR, 165, 180, 185–87, 212, 226
NCR 3125 pen computer, 185–87, 191–94
NEC, 202
Newton. See Apple Newton personal digital assistant (PDA)
New York Times, 150, 163, 178, 242
Next (company), 154–55
NIH ("not invented here"), 156
Nishi, Kay, 67
Non-disclosure agreements (NDAs), 63, 65, 153
"Notebook" metaphor, 153–54; as prototype for next generation of computers, 15; as prototype for pen computers, 24–27, 35–36, 114, 167; user interface, 175
Novell, 255, 265
Numero (Pen Magic), 226–27

Object-oriented programming, 18, 122, 201
"Official Airline Guide," 28
Olivetti Corp., 69, 202
Olsen, Ken, 88

Operating systems, 16, 40, 50, 160, 161, 212, 236, 262–63

Opinions of Counsel, 220–21

Oracle (company), 40

OS/2 (IBM), 117, 122, 128, 153–54, 212, 215–16, 296; Pen OS/2, 216, 296

Ouye, Mike, 38–40, 57, 93, 94, 208–9, 250

Panini, 46

Patriot Partners, 201

PC/Computing, 214

PC Expo, 185, 194

PC Forum, 173–78, 188, 191

PC Magazine, 214

PC Week, 149, 154, 155, 225, 226

PC World, 168

PenApps, 165, 193, 226

Pen computers: advantages of, 30; applications for, 63, 64, 67, 91–92, 101–2, 114, 157, 161, 165, 168, 170–71; demo machines for, 161, 166–68; demos of prototypes for, 60, 61–64, 66–67, 70, 71, 73, 77, 93, 111, 112–14; developing components for, 34–36; developing idea of, 1, 13–18, 50; electronic ink and pens for, 13, 42, 56, 57, 103, 177; hardware development for, 35, 56, 93, 102–3, 107–11, 161, 208–9; as hybrid laptops, 128–29; market potential for, 70, 72, 83, 87, 105, 212, 247; product launch for, 166–68; prototypes for, 34, 51, 56–58, 60, 111; sensing grids for, 103; user testing of, 161; weight of, 34; working prototype for, 109–11

Pen Magic, 226–27

Pen OS/2. *See* OS/2.

Penpoint: awards for, 214, 215, 218; demonstrations of, 166–68, 174–75, 183, 192–93, 214; "developer's release" of, 160–62, 163–69, 224; documentation for, 40, 160, 161, 201, 229; full release of, 224–27; and Hobbit, 217; and Intel, 217; Japanese version, 217, 247, 253; licensing of, 165–66, 188–90, 207, 217, 225–26; logo for, 161–62; press coverage of, 225; pricing of, 180; RISC architecture for, 199–200, 202–3, 206–9; sales projections for, 247, 248, 251; sales to NCR, 185–87; size of, 215; software development for, 35–36, 56, 102, 107, 224, 226–27, 253, 260; source code for, 144–45, 215, 250–52, 260, 293; version 1.0, 226–27; version for customers, 219

Pen-sensing software, 56, 58

Pension funds, 219, 220

Pensoft, 165, 194–95, 292–93

Pen Windows (Windows for Pen Computing), 177, 180, 181, 188, 189, 192–93, 213–16, 223, 235, 237, 296

Perot, Ross, 69

Personal communicators, 232–33, 245, 246, 252, 258, 262

Personal computers (PCs), 6, 14–15, 15

Personal digitial assistants (PDAs), 228, 235, 236

Personal information managers (PIMs), 6–7, 165, 292–93

Personal Software Corp. (Visi-Corp), 11

Phoenix Technologies, 188–91

Pollack, Andy, 178